# FEM for Springs

Springer-Verlag Berlin Heidelberg GmbH

M. Shimoseki · T. Hamano · T. Imaizumi (Eds.)

# FEM for Springs

Translated from
the Japanese original edition published by JSSR

Organized by T. Kuwabara

With 248 Figures and 49 Tables

Springer

**Editors:**

Dr. Masayoshi Shimoseki
Chief Engineer
Mitsubishi Steel MFG. Co., Ltd.
Harumi Park Bldg.
2-22, Harumi 3-Chome, Chuo-Ku
Tokyo, 104-8550 / Japan

Dr. Toshiyuki Imaizumi
Director / General Manager
Technology Office
Chuo Spring Co., Ltd.
43-1 Miyashita Fukuta
Miyoshi-Cho, Nishikamo-Gun
Aichi-Ken, 470-0225 / Japan

Dr. Toshio Hamano
Deputy General Manager
Design Department
Suspension Spring Division
NHK Spring Co., Ltd.
3-10 Fukuura, Kanazawa-Ku
Yokohama, 236-004 / Japan

**Organized by:**

Toshio Kuwabara
Japan Society for Spring Research (JSSR)
Kanda-Mikuracho 12
101-0038 Chiyoda-ku
Tokyo / Japan

ISBN 978-3-642-05505-8 ISBN 978-3-662-05044-6 (eBook)
DOI 10.1007/978-3-662-05044-6

Cataloging-in-Publication Data applied for

Bibliographic information published by Die Deutsche Bibliothek
Die Deutsche Bibliothek lists this publication in the Deutsche Nationalbibliografie;
detailled bibliographic data is available in the Internet at <http://dnb.ddb.de>.



http://www.springer.de


Originally published by Springer-Verlag Berlin Heidelberg New York in 2003
Softcover reprint of the hardcover 1st edition 2003



Typesetting: Camera ready by authors
Cover-design: Medio, Berlin
Printed on acid-free paper 62 / 3020 hu - 5 4 3 2 1 0 -

# Foreword

The Japanese original edition of "FEM for Springs" was published in 1997, to commemorate the 50$^{th}$ anniversary of Japan Society for Spring Research (JSSR). While there have been many books published about Finite Element Method (FEM), this book was among the first to address the application of FEM to spring design.

When asked about springs, one might imagine a mere shape of helical coil. However, there are many more varieties of shapes and functions in the application of springs. Consequently, some are very difficult to calculate by design formula. FEM gives the solutions to those advanced engineering cases.

Nowadays, it is strongly desired to have a design method for springs as a common base from a global point of view. Under these circumstances, JSSR planned to publish an English version of "FEM for Springs". By improving the contents and adding many examples, this book, FEM for Springs, has been brought to completion. It is a truly significant event.

I am confident that this book is suitable for engineers in worldwide industrial sectors and for college students as well.

Kenji Kanazawa

Chairman,

Japan Society for Spring Research

Professor Dr.,

Chuo University

# Preface

I believe this book is the first in the world to describe spring analysis using the finite element method (hereafter FEM). It is well known that FEM has already been applied to a variety of fields, such as the theory of structures, hydrodynamics, and heat transfer, in addition to its applications to aerospace technologies. More recently, its use has spread into more general science and technology areas as a well-understood method of numerical analysis for solving differential equations. It has also taken root in design practice as well as academic research calculation.

In other words, there has been a shift from 'Design by Rule' to 'Design by Analysis' in design area.

FEM has also been actively studied in the Japan Society for Spring Research, but I think this method has not sufficiently or practically penetrated the workplace. Some engineers assert that it is not really required as the existing design method has already been completed. Is this assertion really valid? It is unfair to criticize the FEM without properly understanding it. General mechanical engineers who are unfamiliar with spring technology typically consider spring engineers to be content to stick with existing technology.

Since FEM analysis is all-purpose, it can certainly be applied to springs. As mentioned above, and bearing in mind the past achievements of integrating interdisciplinary variety in the natural sciences, it is very important for general mechanical engineers to understand spring analysis using FEM, and there will be growing demand for FEM to be fully understood by spring engineers as well.

From the very outset, I have striven to make this book of interest to readers in both of the following categories.

Category A readers: Spring engineers aiming to learn about the FEM

Category B readers: General engineers who are unfamiliar with spring technologies

I have set out to attract readers in category A into getting involved in FEM, while also encouraging those in category B to learn about spring technologies.

This book is comprised of both general theories (Chapters 1 ~ 3) and itemized discussions (Chapter 4 onward).

Chapters 1 and 2 are intended to give category A readers an outline of FEM using spring models (called elastic elements in this book). Additionally, increasing awareness among category B readers that spring models as actual springs are somewhat different to what might first be imagined is a secondary aim of Chapter 2. Chapter 3 covers the expected advantages of using FEM on spring technologies and articles concerning limits to the application of this analysis method as a summary of general theory. I have tried to offer simplified explanations, as I would like even those who are not mathematically inclined to at least read this chapter. If such people browse through the parts they are interested in, I will feel I have accomplished something. The central feature for category B readers in this chapter is to understand that springs afford a mine of difficult nonlinear equations. Chapter 4 introduces the representative element types to be used for succeeding chapters, and simply describes points for mesh division. Chapters 5 and 6 introduce a variety of specific analysis examples per spring type. Chapter 5 is mainly comprised of primary analysis examples while Chapter 6 focuses on nonlinear-related issues.

The following basic policies were adopted when selecting examples.

- Analysis shall be undertaken using commercially available software.
- Comparisons to experimental data shall be made as far as possible.
- Research papers shall be introduced for only cited references.
- Distinctions shall be drawn with existing solutions.

Finally, I would like to express my appreciation to the members of the Japan Society for Spring Research for the many examples they provided.

January 2003

Masayoshi Shimoseki
Working Group Leader for 'Finite Element Analysis of Springs'

# Author Index

Text:

Chapter 1; Dr. Masayoshi Shimoseki, Mitsubishi Steel MFG Co., Ltd.
Chapter 2; Dr. Masayoshi Shimoseki, Mitsubishi Steel MFG Co., Ltd.
Chapter 3; Dr. Masayoshi Shimoseki, Mitsubishi Steel MFG Co., Ltd.
Chapter 4; Dr. Toshio Hamano, NHK Spring Co., Ltd.
Chapter 5; Dr. Toshiyuki Imaizumi, Chuo Spring Co., Ltd.
Chapter 6; Dr. Toshio Hamano, NHK Spring Co., Ltd.

Example providers:

Takashi Sato, NHK Spring Co., Ltd.
Satoshi Suzuki, Chuo Spring Co., Ltd.
Koji Goto, Chuo Spring Co., Ltd.
Masahiro Yasui, Chuo Spring Co., Ltd.
Hideto Uruno, Mitsubishi Steel MFG Co., Ltd.
Kanna Chino, Mitsubishi Steel MFG Co., Ltd.

Translation supervisor:

Toshio Kuwabara, JSSR

# Contents

# 1 SPRINGS AND ELASTIC COMPONENT

In the theory of elastic vibration, an analytical model of the spring comes into effect on an assumption that the spring develops stiffness, and its own mass shall be ignored. The stiffness is also known as a spring constant. The definition of spring constant will be discussed in section 1.1. The factor of mass will be treated independently when the system has a mass element. Thus, the image of an abstract spring has been created. Some people might confuse this image with actual springs. In order to prevent such misinterpretation, an image of elastic components will be introduced making a distinction from actual springs in section 1.2. The actual spring is a three-dimensional elastic body having its own mass. For example, a tension spring of a spring balance actually has a stiffness of compression, torsion, bending and shear stiffness together with the inherent stiffness of tension.

*The elastic component* (The elastic component is defined as an idealized one-dimensional elastic body.) ignores an actual three-dimensional elastic body. The elastic component is referred to as a truss, which is the most basic elastic member. By applying the finite element method to the truss (elastic component), basic examples can be demonstrated. In section 1.3 and 1.4, the outline of FEM for elastic components will be presented. Then, FEM for a plane element will follow in section 1.5.

## 1.1 SPRING CONSTANT

### 1.1.1 Definition of the Spring

For the time being we will define the spring as follows: A spring is a mechanical component that can bear larger elastic deformation than other mechanical parts. In some cases the linear relationship between the applied force and the resulting deflection cannot be maintained, even though the force is within the elastic range of the material used for the spring. This type of nonlinearity is called geometrical nonlinearity, as it arises from geometrical conditions, i.e., a large deformation.

Generalization about the nonlinearity in the strength of material will be introduced in section 2.2; only the geometrical nonlinearity in this section will be explained here.

The stress and strain are supposed to keep a linear relationship in an elastic body, however, large deformation sometimes breaks this principle.

In spite of the fact that the geometrical nonlinearity exists, the spring constant — a concept of linearity — is widely accepted. Readers should keep this in mind.

### 1.1.2 Tangential Gradient

Consider an arbitrary one variable equation

$$y = y(u). \tag{1.1}$$

By differentiating (1.1) we obtain

$$dy = \left(\frac{dy}{du}\right) du. \tag{1.2}$$

The differential coefficient $(dy/du)$ represents the tangential gradient and is a function of $u$. This is illustrated in Fig. 1.1.

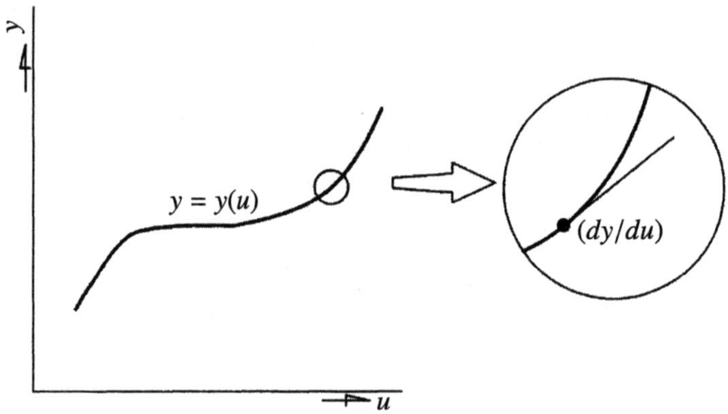


Figure 1.1: Tangential Gradient

Let us assume that Fig. 1.1 is a characteristic diagram of a spring. The concept of spring constant has an inherent meaning only when (1.1) is regarded as a linear equation, i.e., (1.2) is constant. Generally, springs do not have a constant tangential gradient.

### 1.1.3 System of Multiple Degrees of Freedom

When a spring is regarded as a truss with one degree of freedom, the element is an elastic component as was discussed in the introduction of this chapter. The stiffness of the truss $k$ is given as follows;

$$k = \frac{EA}{L}, \tag{1.3}$$

where, $A$ is the cross-sectional area, $L$ is the length and $E$ is Young's modulus.

It is important to be aware that no such truss element actually exists. The truss element is only a mechanical model and the real object is a straight bar. A straight bar is called a truss only when an axial force (tensile or compressive) acts on it. In cases where shearing force or moment is applied to the bar, it is called a beam or a rahmen.

In the same manner, the name of a coiled spring depends on the nature of the applied forces. When axial force is applied, it is a tensional or a compression spring. If moment is applied it is a torsion spring. Consider the example of helical compression springs, they are simultaneously subjected to torsional moment and compressional forces because the change of the pitch angle causes a rotation of the spring wire (which will hereafter be called wire) and the frictional force at the end turn tends to resist the rotation. In addition, the lateral force (shearing force) is necessary for compressing the spring while keeping the plane of the end turns in parallel. It should be understood that pure helical compression springs do not exist and that actual springs are constantly exposed to composite forces.

Based on the above statement, the helical spring should be recognized as a 'spatial curved beam' coiled up in a three dimensional-space[1]. For a beam placed in three-dimensional space there exist three-dimensional forces and moments around three coordinates. This means that there are six degrees of freedom for each point. Unfortunately, this does not mean that our problem can be treated with six proportional equations using six spring constants independently. Our problem is much more complicated because each degree of freedom interacts with each other. A generalized equational expression shall be presented as follows:

$$\boldsymbol{P} = \boldsymbol{K}\boldsymbol{u}, \tag{1.4}$$

where, $\boldsymbol{P}$ is the combined force (including moment) and $\boldsymbol{u}$ the combined displacement (including the rotational angle), $\boldsymbol{K}$ is a cluster of spring components called the stiffness matrix.

(1.4) is based on an assumption that linearity does exist. In the case of nonlinearity, we arrive at our equation by differentiating $\boldsymbol{P}$ and $\boldsymbol{u}$:

$$d\boldsymbol{P} = \frac{d\boldsymbol{P}}{d\boldsymbol{u}} d\boldsymbol{u}. \tag{1.5}$$

The equation (1.5) is similar to (1.2)

# 1.2 ELASTIC COMPONENT IN A VIBRATION SYSTEM

## 1.2.1 Vibration Equations

A linear vibration equation with one degree of freedom, as is shown in Fig. 1.2, is given by

$$m\ddot{u} + c\dot{u} + ku = p, \tag{1.6}$$

where, $m$: Mass, $c$: Damping coefficient, $k$: Spring constant, $p$: External force, $u$: Displacement, $\dot{u}$: Velocity, $\ddot{u}$: Acceleration.

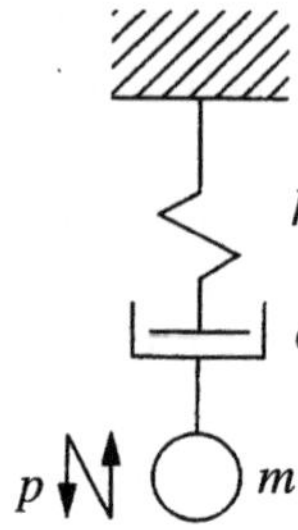


Figure 1.2: *mck* system

In the case of free vibration without damping, we get

$$m\ddot{u} + ku = 0. \tag{1.7}$$

This is an equation of natural vibration. The vibration model is shown in Fig. 1.3.

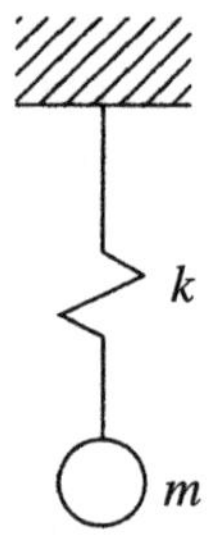


Figure 1.3: *mk* system

This is the simplest case of the vibration. The mass $m$ and the spring constant $k$ are essential elements for a model of vibration. Is it true that a simple spring alone cannot vibrate by itself ? No, the spring constant $k$ without $m$ does not exist actually. The mass is distributed over the spring body itself. Hence a single spring can vibrate by itself and it has its own natural frequency. The cause of such a misunderstanding is the confusion of spring constant $k$, which is no more than an abstract concept of spring, with an actual spring having its own mass.

### 1.2.2 Chords of a Guitar

Consider the vibration of chords of a guitar as an example of the natural vibration of springs. For the bass chords of guitars, metal round wound wires are utilized. This could be a sort of coil springs. The vibration of a plucked chord is a kind of natural vibration.

The treatment of transverse vibration in the coil spring requires an additional explanation in its own way. Instead, an analysis of transverse vibration on a normal chord (not on the round wound chord) will be given in this paragraph. In this example, we do not see any constituting factors corresponding to either $m$ or $k$ as shown in Fig. 1.3. The analysis does contain a dispersing operation of continuous quantity and represents a typical model of the finite element method. Let us assume that the chord travels in the direction of $x$. Next, the total length $L$ is divided into $(n + 1)$ equal parts and each dividing point is numbered as shown in Fig. 1.4. The segmented mass of the chord is concentrated equally at each dividing point.

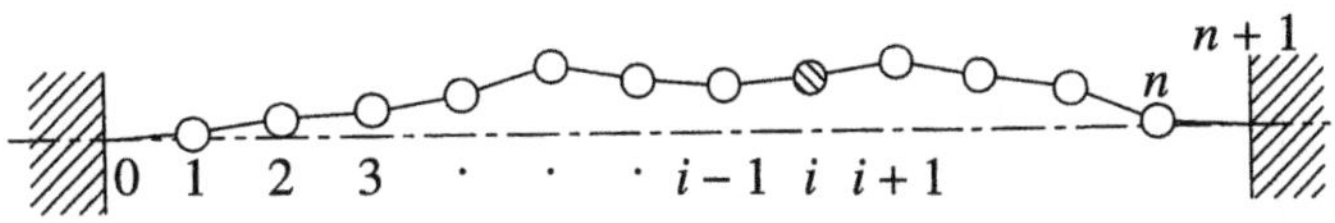


Figure 1.4: Transverse Vibration of a Chord

Let $m$ be the mass of each dividing point, $l$ be the distance between dividing points, and $T$ be the tension of the chord; the transverse displacement $u_i$ in the $y$ direction of the $i$-th dividing point is assumed to be minute. Fig. 1.5 shows the neighborhood of the $i$-th dividing point in enlarged scale.

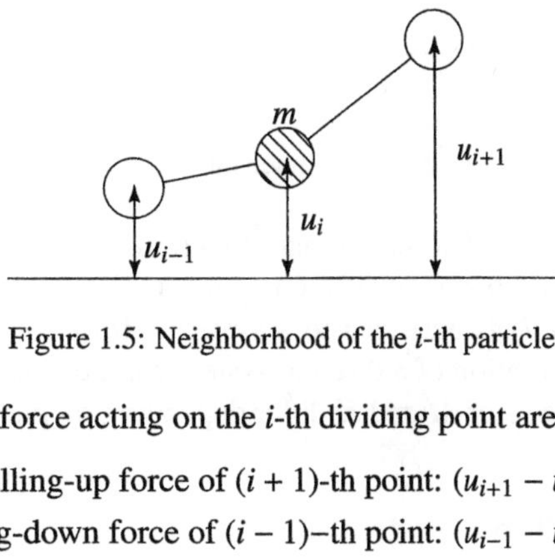


Figure 1.5: Neighborhood of the $i$-th particle

$y$ component of a force acting on the $i$-th dividing point are as follows;

$$\text{pulling-up force of } (i+1)\text{-th point: } (u_{i+1} - u_i)T/l,$$
$$\text{pulling-down force of } (i-1)\text{–th point: } (u_{i-1} - u_i)T/l.$$

By inserting these components into $p$ of the right side of (1.6), we get a differential equation as follows:

$$m\ddot{u}_i = (u_{i+1} - u_i)T/l + (u_{i-1} - u_i)T/l \qquad i = 1, \ldots, n. \tag{1.8}$$

Consider the boundary conditions of both ends

$$i = 1: \; u_{i-1} = u_0 = 0,$$
$$i = n: \; u_{i+1} = u_{n+1} = 0.$$

The equation (1.8) will be expressed in the following matrix equation.

$$\begin{bmatrix} m & 0 & \ldots\ldots & & 0 \\ 0 & m & 0 & \ldots & 0 \\ 0 & 0 & m & 0 \ldots & 0 \\ & \ldots\ldots\ldots & & & \\ 0 & \ldots & & 0 & m \end{bmatrix} \begin{Bmatrix} \ddot{u}_1 \\ \ddot{u}_2 \\ \ddot{u}_3 \\ \ldots \\ \ddot{u}_n \end{Bmatrix} = \frac{T}{l} \begin{bmatrix} -2 & 1 & 0 & & \ldots & 0 \\ 1 & -2 & 1 & 0 & \ldots & 0 \\ 0 & 1 & -2 & 1 & 0 \ldots & 0 \\ & & \ldots\ldots\ldots & & & \\ 0 & & \ldots\ldots & 0 & 1 & -2 \end{bmatrix} \begin{Bmatrix} u_1 \\ u_2 \\ u_3 \\ \ldots \\ u_n \end{Bmatrix}. \quad (1.9)$$

Equation (1.9) can be rearranged into an expression similar to (1.4) to arrive at

$$\boldsymbol{M}\ddot{u} + \boldsymbol{K}u = 0, \quad (1.10)$$

where

$$\boldsymbol{M} = \begin{bmatrix} m & 0 & \ldots\ldots & & 0 \\ 0 & m & 0 & \ldots & 0 \\ 0 & 0 & m & 0 \ldots & 0 \\ & \ldots\ldots\ldots & & & \\ 0 & \ldots & & 0 & m \end{bmatrix}, \quad \boldsymbol{K} = \frac{T}{l} \begin{bmatrix} 2 & -1 & 0 & & \ldots\ldots & 0 \\ -1 & 2 & -1 & 0 & \ldots & 0 \\ 0 & -1 & 2 & -1 & 0 \ldots & 0 \\ & & \ldots\ldots\ldots & & & \\ 0 & & \ldots\ldots & 0 & -1 & 2 \end{bmatrix}.$$

Equation (1.10) is completely the same form as that of the longitudinal vibration of springs aligned in series as shown in Fig. 1.6

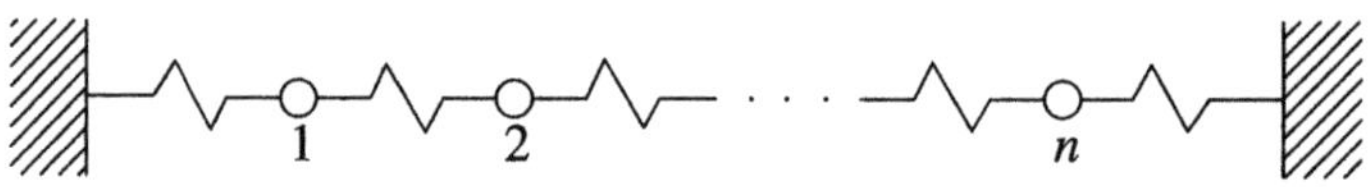


Figure 1.6: Elastic Components Aligned in Series

Please observe that the distributed mass functions as the concentrated mass and the tension force represents the elastic component. As the result, a system of one degree of freedom (1.7) is extended to a system with a multi degree of freedom (1.10). The natural vibration of a discrete system can be expressed by the equation (1.10). Only the contents of $\boldsymbol{M}$ and $\boldsymbol{K}$ differ depending on phenomena.

### 1.2.3 Wave Equation

Equation (1.10) is a formulation of the transverse vibration of a chord using a discrete model. In the case of continuous model, the vibration is expressed by the partial differential equation (1.11).

$$\frac{\partial^2 u}{\partial t^2} = \left(\frac{T}{\rho}\right) \frac{\partial^2 u}{\partial x^2}, \quad (1.11)$$

where $\rho$ is the linear density ($= m/dx$).

Replacing $l$ in (1.8) with $\Delta x$, we get

$$\ddot{u}_i = T/m\left(\frac{u_{i-1} - 2u_i + u_{i+1}}{\Delta x^2}\right)\Delta x.$$

Taking the limit $\Delta x \to 0$, we get

$$\frac{\partial^2 u}{\partial u^2} = \left(\frac{T}{m}\right)\frac{\partial^2 u}{\partial x^2}\,dx.$$

Introducing the linear density $\rho$, we can obtain (1.11).

Equations (1.10) and (1.11) treat the same phenomenon differently. The greatest point of difference is that one is an ordinary differential equation and the other, a partial differential equation. In (1.11), the displacement is treated as the two variable function of $x$ and $t$. In (1.10), the displacement is expressed only as the function of $t$ by dispersing $x$. The partial differential equation is simplified into the ordinary differential equation through a spatially discrete treatment. (1.10) is expressed as $n$ dimensional simultaneous equation. $n$ dividing points of $x$ serves this simplification. The treatment of time $t$ will be described later in section 2.3.

## 1.3 FIRST STAGES OF ANALYSIS

### 1.3.1 Orientation

There are two important points in the finite element method, which are the handling of the boundary condition and the construction of the equation from the element to the global structure. The conception of the finite element method differs significantly from the conventional analytical method as a result of these two points.

**(1) Handling of Boundary Condition** In the case of a guitar chord analysis, the boundary condition is given in advance. However, it is not always useful to establish boundary conditions in the beginning. In the design process, it is often necessary to consider a problem from different angles and with various boundary conditions. In the design of an oblique suspension bridge for instance, it is necessary to take every possible case into consideration such as, vehicle traffic, strong winds and earthquakes. The conventional method incorporating the boundary condition beforehand, makes it necessary to repeat the analysis by changing the setting of conditions. This method is impractical and time consuming for analysis requiring diversified examinations. By using the finite element method, only the mathematical expression representing the mechanical property of the structure is determined in advance and the problem is analyzed by specifying the boundary condition afterward. Using this method, changing the boundary condition is easy to accomplish. It is only necessary in the latter half of the analysis.

**(2) From Element to Global** The finite element method assumes that the elastic body is an assembly of elastic components. In an multiple component model, the combined spring constant can be derived from each spring constant.

In the finite element method, where the domain of analysis is divided into a limited number of sub-domains, the global stiffness is mechanically constructed by combining the properties of sub-domains (elements) as in the case of the spring constant.

### 1.3.2 Steps of Analysis

The following chart shows the analytical steps of the finite element method in comparison with the conventional way (General elastic body). A structure constructed by the combination of elastic components is referred to as a 'component structure' in this book.

| | [General elastic body] | [Component structure] |
|---|---|---|
| Step 1 | Disassemble into each element<br>Find element stiffness | Disassemble into each component<br>Find each spring constant |
| | ↓ | ↓ |
| Step 2 | Join elements<br>Find global stiffness | Join components<br>Find composite spring constant |
| | ↓ | ↓ |
| Step 3 | Give boundary conditions | Give boundary conditions |
| | ↓ | ↓ |
| Step 4 | Solve equation | Solve equation |

To solve with different boundary conditions, return to Step 3 and repeat the process again.

Please note the resemblance on each occasion. Step 3 and Step 4 are exactly the same. Step 1 and Step 2 do not differ substantially even if their expressions are different. Furthermore, any partial amendment of their shape is readily carried out by replacing the element stiffness of the concerned position, similar to the way replacing a defective mechanical part. The same principle applies to anything from an electromagnetic field analysis to a fluid dynamic analysis. The necessary procedure is to insert each characteristic value instead of the stiffness (spring constant). In short, the analyzing process holds itself unchanged even if the nature of the problem is different.

In the case of a linear problem, Step 4 results in simultaneous linear equations. In the finite element method, it is a common practice to solve sparse (those with a lot of zero components) and a huge number of simultaneous equations.

For the analysis of the nonlinear problem, the incremental method, i.e. a method to solve segmented linear analysis, is used. Concerning these mathematical equations, many practical solutions have been developed. These topics belong to the field of applied mathematics and are beyond the scope of this book. The outline of such techniques will be described further in section 2. For more detailed study, refer to [2].

### 1.3.3 A Pitfall in the Approximate Solution

To go into a little more detail concerning the previously mentioned steps, the elastic body and the component structure differ in their contents of Step 1. A mesh of components is, after all, simply an aggregate of discrete elastic components, and the number or the combination of components constituting the aggregate is inevitably fixed. By contrast in the case of a continuous body, there exists a degree of arbitrariness in dividing it into elements. There arise individual differences in the ways to divide the continuous domain, and these differences in the dividing procedure may eventually influence the result of the analysis. In any case, while Step 1 is an exact solution for the component structure, it is only an approximate solution for the elastic body.

When an external force is applied to an elastic body, a displacement field is generated within the domain. If the domain has a complicated configuration, then we assume the displacement field will also bear a complex structure. Even in instances where the configuration is complicated, it should be possible to make an approximation with equations of lower order, provided that the discretization of the domain is sufficiently small. Fig. 1.7 illustrates the difference in the discretization in the case of a one dimensional domain.

As the problem is one dimensional, the domain is represented by $a < x < b$ on the $x$ axis and the divided part of the domain, (line segment), is the element domain.

The function of $x$, $u(x)$, is represented by a solid line. The approximation by linear function when the domain is divided into three equal elements is illustrated by a dot-dashed line and a broken line for division into unequal parts. Thus, the accuracy of approximation can be improved by changing the element size fit for the functional value.

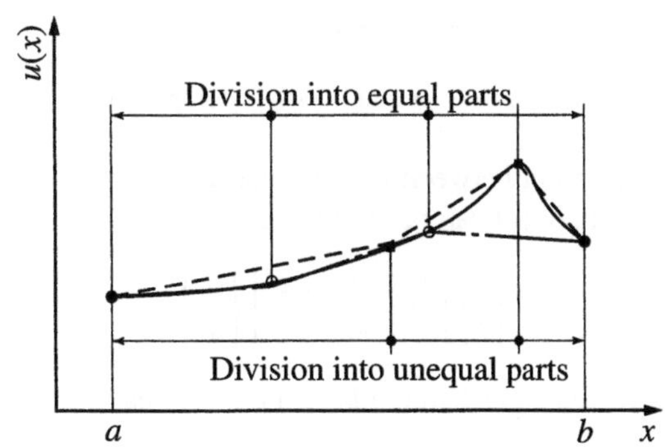


Figure 1.7: Difference by Division

In Fig. 1.7, the approximation was undertaken with linear equations. The accuracy of approximation should be improved when the domain is approximated with quadratic equations: the accuracy of approximation is expected to improve by increasing the order of approximate equation. In order to obtain a designated accuracy, the number of divisions can be reduced through the use of higher order equations. Development of this idea eventually results in establishing the selecting criterion of the element type. As will be introduced in chapter 4, various finite elements have been invented so far. With this in mind, it becomes important to select the element most appropriate to the type of problem.

The above consideration shows that the accuracy of the solution changes in accordance to the method of dividing, or type of element in the case of an elastic body. In the case of elastic components, on the other hand, the displacement function should be expressed by the exact linear equation because the deformation is uniform. In other words, the stiffness of the element obtained by Step 1 is an exact solution in the case of elastic components, but an approximate solution in the case of an elastic body.

### 1.3.4 Element Stiffness of Elastic Component

Let assume that an elastic component $k$ is placed parallel to the $x$ axis, external forces act only in the $x$ direction and therefore the displacement takes place only in the direction of $x$. Both ends of the elastic component are given nodal numbers as shown in Fig. 1.8. The external force $P$ and the displacement $u$ are distinguished by attaching nodal numbers.

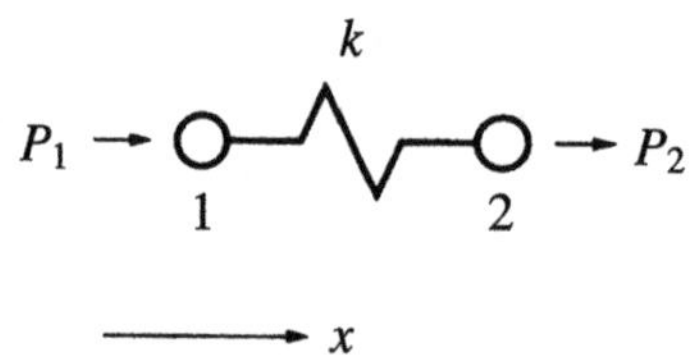


Figure 1.8: A single elastic component

In this case, the relation between the external force and the displacement concerning the elastic component can be expressed as

$$\begin{Bmatrix} P_1 \\ P_2 \end{Bmatrix} = \begin{bmatrix} k & -k \\ -k & k \end{bmatrix} \begin{Bmatrix} u_1 \\ u_2 \end{Bmatrix}. \tag{1.12}$$

The first segment of the right hand side of (1.12) is called the element stiffness matrix.

Equation (1.12) indicates the equilibrium equation in a floating state, since no boundary condition is included, and it can deal with any boundary condition. The proof of (1.12) is as follows:

**(1) Castigliano's theorem** As the elongation of the elastic component is $u_2 - u_1$, the strain energy $U$ stored in the component becomes

$$U = \frac{1}{2}k(u_2 - u_1)^2,$$

therefore, we get

$$P_1 = \frac{\partial U}{\partial u_1} = k(u_1 - u_2),$$
$$P_2 = \frac{\partial U}{\partial u_2} = k(-u_1 + u_2).$$

**(2) Principle of virtual work** The elastic component shown in Fig. 1.8 is in a static equilibrium and the mechanical quantities are as follows:

$$\text{External force on the node 1} = P_1,$$
$$\text{External force on the node 2} = P_2,$$
$$\text{Axial force} = k(u_2 - u_1).$$

Keeping the above mechanical quantities in equilibrium, let the virtual displacements shown below take place.

$$\text{Displacement of the node 1} = \delta u_1,$$
$$\text{Displacement of the node 2} = \delta u_2,$$
$$\text{Contraction (or elongation)} = \delta u_2 - \delta u_1.$$

We then arrive at the virtual works.

$$\text{Internal virtual work} = k(u_2 - u_1) \cdot (\delta u_2 - \delta u_1),$$
$$\text{External virtual work} = P_1 \delta u_1 + P_2 \delta u_2.$$

Equating both the internal and external virtual works, we get

$$\{k(u_1 - u_2) - P_1\}\delta u_1 + \{k(-u_1 + u_2) - P_2\}\delta u_2 = 0.$$

This formula stands under any value of the virtual displacement $\delta u_1$ and $\delta u_2$, and we once again get (1.12).

## 1.4 ONE-DIMENSIONAL COMBINATION OF COMPONENTS

### 1.4.1 Coupling Between Components

Consider two elastic components $a$ and $b$. Each end of the component has the nodal number as is shown in Fig. 1.9.

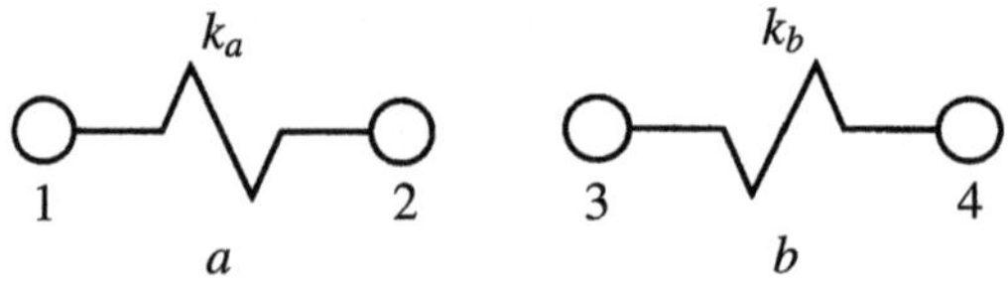


Figure 1.9: Two elastic components

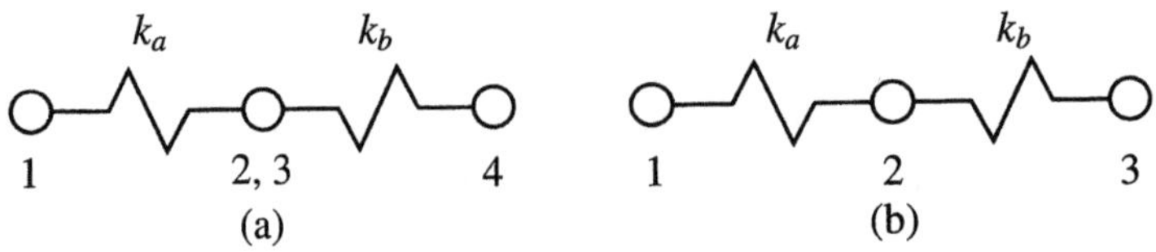


Figure 1.10: Elastic components aligned in series

From (1.12), the element stiffness is given as

$$\begin{Bmatrix} P_1 \\ P_2 \end{Bmatrix} = \begin{bmatrix} k_a & -k_a \\ -k_a & k_a \end{bmatrix} \begin{Bmatrix} u_1 \\ u_2 \end{Bmatrix}, \tag{1.13}$$

$$\begin{Bmatrix} P_3 \\ P_4 \end{Bmatrix} = \begin{bmatrix} k_b & -k_b \\ -k_b & k_b \end{bmatrix} \begin{Bmatrix} u_3 \\ u_4 \end{Bmatrix}. \tag{1.14}$$

Assembling these equations, we get the following expression

$$\begin{Bmatrix} P_1 \\ P_2 \\ P_3 \\ P_4 \end{Bmatrix} = \begin{bmatrix} k_a & -k_a & 0 & 0 \\ -k_a & k_a & 0 & 0 \\ 0 & 0 & k_b & -k_b \\ 0 & 0 & -k_b & k_b \end{bmatrix} \begin{Bmatrix} u_1 \\ u_2 \\ u_3 \\ u_4 \end{Bmatrix}. \tag{1.15}$$

In the case of Fig. 1.10 (a), where the node 2 of $a$ and the node 3 of $b$ are connected, a relation between two nodes shall be considered as follows.

$$\text{Displacement: } u_2 = u_3. \tag{1.16}$$

If no external forces act on the coupled nodal point, we get a balance of force:

$$\text{Force: } P_2 + P_3 = 0. \tag{1.17}$$

Next, we try to incorporate (1.16) and (1.17) into (1.15). Since $u_2$ equals $u_3$, we eliminate one of the two displacements ($u_3$). The 3rd column figures shall be added to 2nd column in the coefficient matrix of equation (1.15). This results in:

$$\begin{Bmatrix} P_1 \\ P_2 \\ P_3 \\ P_4 \end{Bmatrix} = \begin{bmatrix} k_a & -k_a & 0 \\ -k_a & k_a & 0 \\ 0 & k_b & -k_b \\ 0 & -k_b & k_b \end{bmatrix} \begin{Bmatrix} u_1 \\ u_2 \\ u_4 \end{Bmatrix}. \tag{1.18}$$

Next, we incorporate (1.17) into (1.18); add the third row to the second row of the coefficient matrix. After the above operation, the node can be renamed from 4 to 3 as is shown in Fig. 1.10(b). To put the mathematical expression in order, we get:

$$\begin{Bmatrix} P_1 \\ 0 \\ P_3 \end{Bmatrix} = \begin{bmatrix} k_a & -k_a & 0 \\ -k_a & k_a + k_b & -k_b \\ 0 & -k_b & k_b \end{bmatrix} \begin{Bmatrix} u_1 \\ u_2 \\ u_3 \end{Bmatrix}. \tag{1.19}$$

### 1.4.2 Generalized Matrix Equation for Coupled Elastic Components

Suppose the joining point receives not only a reaction from the elastic components but also from an external force. Then, three forces acting on renamed node 2 as shown in Fig. 1.11 are as follows:

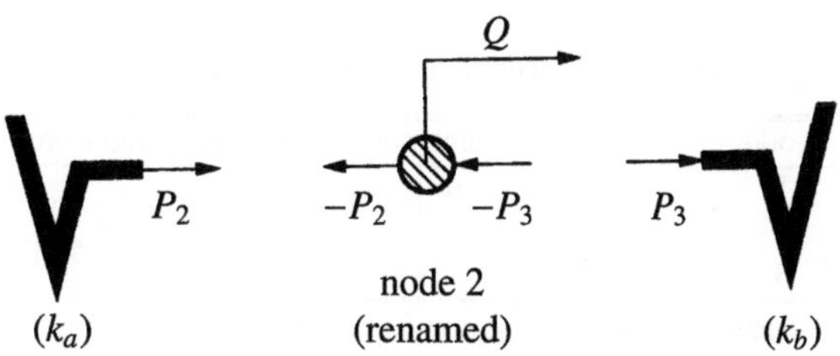


Figure 1.11: Equilibrium of forces at a joining point

Reaction from $k_a$: $-P_2$(inverse sign to $P_2$),
Reaction from $k_b$: $-P_3$(inverse sign to $P_3$),
External force on the node: $Q$.

The sum of these forces must be zero because the joining point does not move. Then we get

$$-P_2 - P_3 + Q = 0. \tag{1.20}$$

The equation (1.17) is a specific case of (1.20) when $Q = 0$. As a rule, $Q$ shall be inserted to the left side of (1.19).

Renaming the external force $Q$ with $P_2$, we get an expression

$$\begin{Bmatrix} P_1 \\ P_2 \\ P_1 \end{Bmatrix} = \begin{bmatrix} k_a & -k_a & 0 \\ -k_a & k_a + k_b & -k_b \\ 0 & -k_b & k_b \end{bmatrix} \begin{Bmatrix} u_1 \\ u_2 \\ u_1 \end{Bmatrix}. \tag{1.21}$$

This coefficient matrix refers to the stiffness of a unit in which two elastic components $a$ and $b$ are joined in a column arrangement. This kind of coefficient matrix representing the stiffness of a global system is called the global stiffness matrix in contrast to the element stiffness.

### 1.4.3 Verification of Boundary Condition Type

In this paragraph, we will see how a boundary condition affects a stiffness matrix by solving (1.21) on two boundary conditions.

**(1) One End Fixed Condition:** Assume a case as shown in Fig. 1.12.

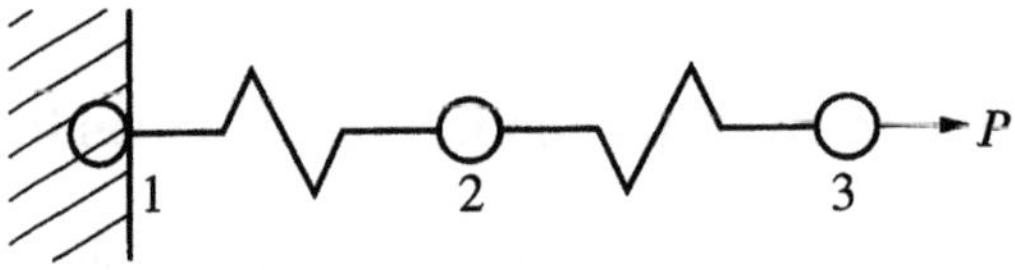


Figure 1.12: A case fixed at one end

The boundary condition is given as shown in Table 1.1.

Table 1.1: Boundary condition of Fig. 1.12

| Node | Displacement | External Force |
|---|---|---|
| 1 | 0 | – |
| 2 | – | 0 |
| 3 | – | $P$ |

Remarks: Dash (–) means an unknown quantity.

Substituting the parameter in (1.21) for the known quantities in Table 1.1, we get:

$$\begin{Bmatrix} P_1 \\ 0 \\ P \end{Bmatrix} = \begin{bmatrix} k_a & -k_a & 0 \\ -k_a & k_a + k_b & -k_b \\ 0 & -k_b & k_b \end{bmatrix} \begin{Bmatrix} 0 \\ u_2 \\ u_1 \end{Bmatrix}. \tag{1.22}$$

Extracting two equations concerning the nodes 2 and 3, it becomes

$$\begin{Bmatrix} 0 \\ P \end{Bmatrix} = \begin{bmatrix} -k_a & k_a + k_b & -k_b \\ 0 & -k_b & k_b \end{bmatrix} \begin{Bmatrix} 0 \\ u_2 \\ u_3 \end{Bmatrix} = \begin{bmatrix} k_a + k_b & -k_b \\ -k_b & k_b \end{bmatrix} \begin{Bmatrix} u_2 \\ u_3 \end{Bmatrix}. \tag{1.23}$$

Solving this equation, we obtain

$$\begin{aligned} \begin{Bmatrix} u_2 \\ u_3 \end{Bmatrix} &= \begin{bmatrix} k_a + k_b & -k_b \\ -k_b & k_b \end{bmatrix}^{-1} \begin{Bmatrix} 0 \\ P \end{Bmatrix} = \frac{1}{(k_a + k_b)k_b - k_b^2} \begin{bmatrix} k_b & k_b \\ k_b & k_a + k_b \end{bmatrix} \begin{Bmatrix} 0 \\ P \end{Bmatrix} \\ &= \frac{P}{k_a k_b} \begin{Bmatrix} k_b \\ k_a + k_b \end{Bmatrix} = P \begin{Bmatrix} k_a^{-1} \\ k_a^{-1} + k_b^{-1} \end{Bmatrix}. \end{aligned} \tag{1.24}$$

Substituting the result of (1.24) into the first row of (1.22), we obtain the unknown reaction at node 1

$$P_1 = \{\, k_a \;\; -k_a \;\; 0 \,\} \begin{Bmatrix} 0 \\ u_2 \\ u_3 \end{Bmatrix} = -k_a u_2 = -P. \tag{1.25}$$

**(2) Both Ends Fixed Condition:** Assume the case as is shown in Fig. 1.13.

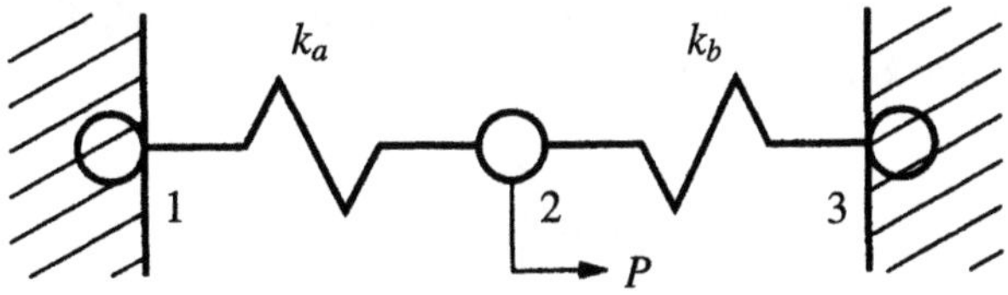


Figure 1.13: A case fixed at both ends

The boundary condition is given in Table 1.2.

Table 1.2: Boundray condition in Fig. 1.13

| Node | Displacement | External Force |
|---|---|---|
| 1 | 0 | – |
| 2 | – | $P$ |
| 3 | 0 | – |

Substituting the parameter in (1.21) for the known quantity in Table 1.2, we get

$$\begin{Bmatrix} P_1 \\ P \\ P_3 \end{Bmatrix} = \begin{bmatrix} k_a & -k_b & 0 \\ -k_a & k_a + k_b & -k_b \\ 0 & -k_b & k_b \end{bmatrix} \begin{Bmatrix} 0 \\ u_2 \\ 0 \end{Bmatrix}. \tag{1.26}$$

Extracting the equation concerning the node 2, we get

$$P = \{\ -k_a \quad k_a + k_b \quad -k_b\ \} \begin{Bmatrix} 0 \\ u_2 \\ 0 \end{Bmatrix} = (k_a + k_b)u_2. \tag{1.27}$$

Substituting the result of (1.27) into the first and third rows of (1.26), we can obtain the unknown reaction at the nodes 1 and 3

$$\begin{Bmatrix} P_1 \\ P_3 \end{Bmatrix} = \begin{bmatrix} k_a & -k_a & 0 \\ 0 & -k_b & k_b \end{bmatrix} \begin{Bmatrix} 0 \\ u_2 \\ 0 \end{Bmatrix} = -u_2 \begin{Bmatrix} k_a \\ k_b \end{Bmatrix} = \frac{-P}{k_a + k_b} \begin{Bmatrix} k_a \\ k_b \end{Bmatrix}. \tag{1.28}$$

As can be understood from the above two examples, an assembly of two elastic components which is joined in a column arrangement, could demonstrate two different solutions depending on their boundary condition. The essential point is to define the displacement and force applied to each component.

### 1.4.4 Parallel Coupling of Elastic Components

In the case of Fig. 1.14, where $k_a$ and $k_b$ are coupled in parallel, displacements are common for $k_a$ and $k_b$ at each end.

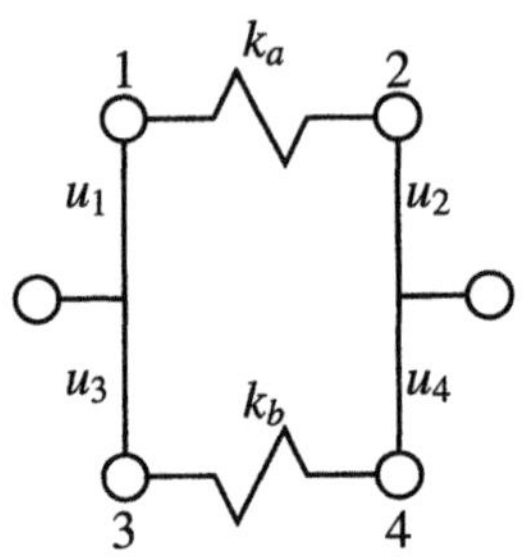


Figure 1.14: Parallel Coupling

Since $u_1 = u_3$ and $u_2 = u_4$, let us consider $u_1$ and $u_2$ as representations of displacement. Adding the third and fourth columns in the coefficient matrix of (1.15) to the first and second columns, we get

$$\begin{Bmatrix} P_1 \\ P_2 \\ P_3 \\ P_4 \end{Bmatrix} = \begin{bmatrix} k_a & -k_a \\ -k_a & k_a \\ k_b & -k_b \\ -k_b & k_b \end{bmatrix} \begin{Bmatrix} u_1 \\ u_2 \end{Bmatrix}. \tag{1.29}$$

Next, the third and fourth rows are added to the first and second rows, correspondingly, considering the combined forces at the nodes.

We get

$$\begin{Bmatrix} P_1 + P_3 \\ P_2 + P_4 \end{Bmatrix} = \begin{bmatrix} (k_a + k_b) & -(k_a + k_b) \\ -(k_a + k_b) & (k_a + k_b) \end{bmatrix} \begin{Bmatrix} u_1 \\ u_2 \end{Bmatrix}. \tag{1.30}$$

Changing the notation of joining points to nodes 1 and 2 as is shown in Fig. 1.15, results in the following expressions:

External force at the new node 1 = External force at the old node 1
+External force at the old node 3,

External force at the new node 2 = External force at the old node 2
+External force at the old node 4.

The equation (1.30) can be expressed finally as

$$\begin{Bmatrix} P_1 \\ P_2 \end{Bmatrix} = (k_a + k_b) \begin{bmatrix} 1 & -1 \\ -1 & 1 \end{bmatrix} \begin{Bmatrix} u_1 \\ u_2 \end{Bmatrix}. \tag{1.31}$$

The coefficient in the above expression is identical to the element stiffness $(k_a + k_b)$ of a single component.

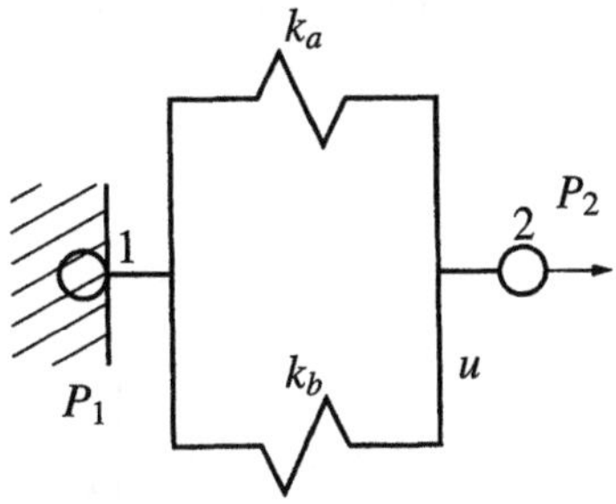


Figure 1.15: One Fixed End

On the boundary condition in Fig 1.15, we obtain

$$\begin{Bmatrix} P_1 \\ P_2 \end{Bmatrix} = (k_a + k_b) \begin{bmatrix} 1 & -1 \\ -1 & 1 \end{bmatrix} \begin{Bmatrix} 0 \\ u \end{Bmatrix} = (k_a + k_b) \begin{Bmatrix} -u \\ u \end{Bmatrix}. \tag{1.32}$$

The operation is carried out with the same procedure irrespective of the type of arrangement or series/parallel coupling of elastic components. Basically, we obtain the global stiffness by taking the related equations of the displacement and the force into account and undertaking mechanical operations such as the rearrangement or addition of the coefficient matrix. Such a treating method is called the direct stiffness method and is well suited for use with computers. The direct stiffness method plays an active role in the finite element method.

### 1.4.5 Transverse Stiffness of Elastic Components

In this paragraph, we will try to construct a more realistic model of the elastic component. As was previously mentioned, an actual helical spring will deform in a lateral direction when loaded laterally. This means that the actual helical spring has stiffness not only in the longitudinal direction but also in the lateral direction. Although the helical spring should be treated as having six-degrees of freedom for a rigorous study, a simplified model with only two-degrees of freedom will be studied in this paragraph. A full-scale analysis will be explained in chapter 5.

Using the notations shown in Table 1.3 and formula (1.12) derived in 1.3.4, we get the following expression for the stiffness:

$$\begin{Bmatrix} P_1 \\ Q_1 \\ P_2 \\ Q_2 \end{Bmatrix} = \begin{bmatrix} k & 0 & -k & 0 \\ 0 & \kappa & 0 & -\kappa \\ -k & 0 & k & 0 \\ 0 & -\kappa & 0 & \kappa \end{bmatrix} \begin{Bmatrix} u_1 \\ v_1 \\ u_2 \\ v_2 \end{Bmatrix}. \tag{1.33}$$

Table 1.3: A case with two-degrees of freedom

| | $x$ direction | $y$ direction |
|---|---|---|
| Force | $P$ | $Q$ |
| Displacement | $u$ | $v$ |
| Spring Constant | $k$ | $\kappa$ |

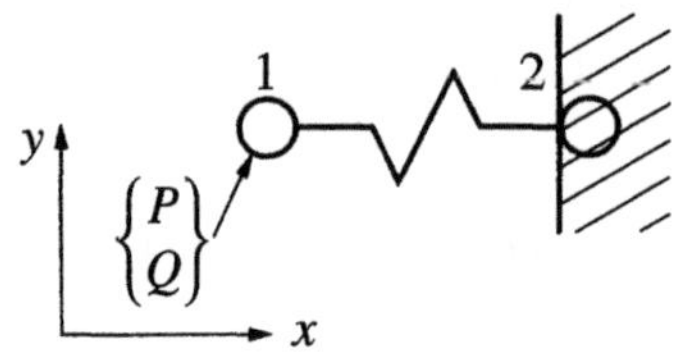


Figure 1.16: An elastic component receiving an oblique force

Table 1.4: Boundray condition in Fig. 1.16

| Node | Displacement | | Force | |
|---|---|---|---|---|
| | $x$ | $y$ | $x$ | $y$ |
| 1 | – | – | $P$ | $Q$ |
| 2 | 0 | 0 | – | – |

Providing the boundary conditions shown in Fig. 1.16 and Table 1.4, we get

$$\begin{Bmatrix} P \\ Q \\ P_2 \\ Q_2 \end{Bmatrix} = \begin{bmatrix} k & 0 & -k & 0 \\ 0 & \kappa & 0 & -\kappa \\ -k & 0 & k & 0 \\ 0 & -\kappa & 0 & \kappa \end{bmatrix} \begin{Bmatrix} u_1 \\ v_1 \\ 0 \\ 0 \end{Bmatrix}. \tag{1.34}$$

Solving the unknown displacements of node 1, where the forces are known, gives us:

$$\begin{Bmatrix} u_1 \\ v_1 \end{Bmatrix} = \begin{bmatrix} k & 0 \\ 0 & \kappa \end{bmatrix}^{-1} \begin{Bmatrix} P \\ Q \end{Bmatrix} = \frac{1}{k\kappa} \begin{Bmatrix} \kappa P \\ kQ \end{Bmatrix} = \begin{Bmatrix} P/k \\ Q/\kappa \end{Bmatrix}. \tag{1.35}$$

Substituting the obtained result into the second and third row of (1.33), we get the reaction force as follows:

$$\begin{Bmatrix} P_2 \\ Q_2 \end{Bmatrix} = \begin{bmatrix} -k & 0 \\ 0 & -\kappa \end{bmatrix} \begin{Bmatrix} u_1 \\ v_1 \end{Bmatrix} = \begin{Bmatrix} -P \\ -Q \end{Bmatrix}. \tag{1.36}$$

In the above case, the stiffness of $x$ and $y$ directions are entirely independent of each other. Therefore the result is self-explanatory.

## 1.5 PLANE STRUCTURES

### 1.5.1 Transformation of Coordinates

Suppose an elastic component is oriented at angle $\theta$ from $x$ axis, as is the case in Fig. 1.17.

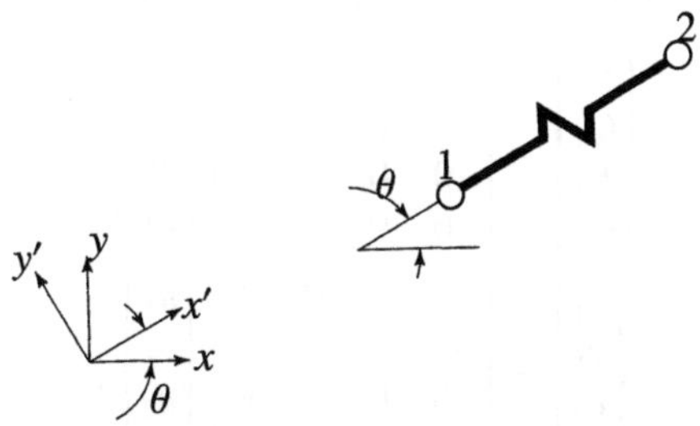


Figure 1.17: A diagonally placed elastic component

Let's assume that the degree of freedom of the component is allowed only in the longitudinal direction and its stiffness is $k$. Although the degree of freedom of the component itself is only one, it possesses stiffness in the $x$ and $y$ directions, separately, as the longitudinal direction and the coordinate axis are not parallel. We call this $x$-$y$ coordinate, the global coordinates, and give the notations as shown in Table 1.5.

Table 1.5: Notations in Fig. 1.17

| | Global coordinate system | | Local coordinate system | |
|---|---|---|---|---|
| | $x$ | $y$ | $x'$ | $y'$ |
| Force | $P$ | $Q$ | $P'$ | $Q'$ |
| Displacement | $u$ | $v$ | $u'$ | $v'$ |

As in the case of (1.33), the equilibrium equation of the component can be expressed as

$$\begin{Bmatrix} P_1 \\ Q_1 \\ P_2 \\ Q_2 \end{Bmatrix} = \begin{bmatrix} k_{11} & k_{12} & k_{13} & k_{14} \\ k_{21} & k_{22} & k_{23} & k_{24} \\ k_{31} & k_{32} & k_{33} & k_{34} \\ k_{41} & k_{42} & k_{43} & k_{44} \end{bmatrix} \begin{Bmatrix} u_1 \\ v_1 \\ u_2 \\ v_2 \end{Bmatrix}. \tag{1.37}$$

Now, the problem is to describe the coefficients of $k_{11}, \ldots, k_{44}$ by $k$ and $\theta$.

Now, we set the $x'$ axis of the local coordinate system in parallel to the component as is shown in Fig. 1.17. The local coordinate system is rotated from the global coordinate system at an angle of $\theta$. Since the component has no stiffness in the $y'$

direction in this local coordinate system, we get from (1.33):

$$\begin{Bmatrix} P'_1 \\ Q'_1 \\ P'_2 \\ Q'_2 \end{Bmatrix} = \begin{bmatrix} k & 0 & -k & 0 \\ 0 & 0 & 0 & 0 \\ -k & 0 & k & 0 \\ 0 & 0 & 0 & 0 \end{bmatrix} \begin{Bmatrix} u'_1 \\ v'_1 \\ u'_2 \\ v'_2 \end{Bmatrix}. \tag{1.38}$$

Considering that the force $(P, Q)$ and the displacement $(u, v)$ are vectors, we get

$$\begin{Bmatrix} P'_1 \\ Q'_1 \\ P'_2 \\ Q'_2 \end{Bmatrix} = \begin{bmatrix} c & s & 0 & 0 \\ -s & c & 0 & 0 \\ 0 & 0 & c & s \\ 0 & 0 & -s & c \end{bmatrix} \begin{Bmatrix} P_1 \\ Q_1 \\ P_2 \\ Q_2 \end{Bmatrix}, \tag{1.39}$$

$$\begin{Bmatrix} u'_1 \\ v'_1 \\ u'_2 \\ v'_2 \end{Bmatrix} = \begin{bmatrix} c & s & 0 & 0 \\ -s & c & 0 & 0 \\ 0 & 0 & c & s \\ 0 & 0 & -s & c \end{bmatrix} \begin{Bmatrix} u_1 \\ v_1 \\ u_2 \\ v_2 \end{Bmatrix}, \tag{1.40}$$

where, $s = \sin\theta$, and $c = \cos\theta$.

The coefficient matrixes of (1.39) and (1.40) are called the coordinate transformation matrixes. Substituting (1.39) and (1.40) into (1.38), we get

$$\begin{Bmatrix} P'_1 \\ Q'_1 \\ P'_2 \\ Q'_2 \end{Bmatrix} = \begin{bmatrix} c & s & 0 & 0 \\ -s & c & 0 & 0 \\ 0 & 0 & c & s \\ 0 & 0 & -s & c \end{bmatrix} \begin{Bmatrix} P_1 \\ Q_1 \\ P_2 \\ Q_2 \end{Bmatrix}$$

$$= \begin{bmatrix} k & 0 & -k & 0 \\ 0 & 0 & 0 & 0 \\ -k & 0 & k & 0 \\ 0 & 0 & 0 & 0 \end{bmatrix} \begin{bmatrix} c & s & 0 & 0 \\ -s & c & 0 & 0 \\ 0 & 0 & c & s \\ 0 & 0 & -s & c \end{bmatrix} \begin{Bmatrix} u_1 \\ v_1 \\ u_2 \\ v_2 \end{Bmatrix}. \tag{1.41}$$

Inverse matrix of the coordinate transformation matrix:

$$\begin{bmatrix} c & s & 0 & 0 \\ -s & c & 0 & 0 \\ 0 & 0 & c & s \\ 0 & 0 & -s & c \end{bmatrix}^{-1} = \begin{bmatrix} c & -s & 0 & 0 \\ s & c & 0 & 0 \\ 0 & 0 & c & -s \\ 0 & 0 & s & c \end{bmatrix}.$$

By multiplying this inverse matrix to the both sides of (1.41) from left side, we get:

$$\begin{Bmatrix} P_1 \\ Q_1 \\ P_2 \\ Q_2 \end{Bmatrix} = \begin{bmatrix} c & -s & 0 & 0 \\ s & c & 0 & 0 \\ 0 & 0 & c & -s \\ 0 & 0 & s & c \end{bmatrix} \begin{bmatrix} k & 0 & -k & 0 \\ 0 & 0 & 0 & 0 \\ -k & 0 & k & 0 \\ 0 & 0 & 0 & 0 \end{bmatrix} \begin{bmatrix} c & s & 0 & 0 \\ -s & c & 0 & 0 \\ 0 & 0 & c & s \\ 0 & 0 & -s & c \end{bmatrix} \begin{Bmatrix} u_1 \\ v_1 \\ u_2 \\ v_2 \end{Bmatrix}$$

$$= k \begin{bmatrix} c^2 & sc & -c^2 & -sc \\ sc & s^2 & -sc & -s^2 \\ -c^2 & -sc & c^2 & sc \\ -sc & -s^2 & sc & s^2 \end{bmatrix} \begin{Bmatrix} u_1 \\ v_1 \\ u_2 \\ v_2 \end{Bmatrix}. \tag{1.42}$$

Now we have the concrete components of (1.37).

To generate a clearer picture, we introduce the following new notations:

$$\boldsymbol{P}=\begin{Bmatrix}P_1\\Q_1\\P_2\\Q_2\end{Bmatrix},\quad \boldsymbol{P}'=\begin{Bmatrix}P_1'\\Q_1'\\P_2'\\Q_2'\end{Bmatrix},\quad \boldsymbol{u}=\begin{Bmatrix}u_1\\v_1\\u_2\\v_2\end{Bmatrix},\quad \boldsymbol{u}'=\begin{Bmatrix}u_1'\\v_1'\\u_2'\\v_2'\end{Bmatrix},$$

$$\boldsymbol{k}=\begin{bmatrix}k_{11}&k_{12}&k_{13}&k_{14}\\k_{21}&k_{22}&k_{23}&k_{24}\\k_{31}&k_{32}&k_{33}&k_{34}\\k_{41}&k_{42}&k_{43}&k_{44}\end{bmatrix},\quad \boldsymbol{k}'=\begin{bmatrix}k&0&-k&0\\0&0&0&0\\-k&0&k&0\\0&0&0&0\end{bmatrix},\quad \boldsymbol{T}\begin{bmatrix}c&s&0&0\\-s&c&0&0\\0&0&c&s\\0&0&-s&c\end{bmatrix}.$$

The process of derivation pursued thus far can be interpreted as follows: $\boldsymbol{P}'=\boldsymbol{T}\boldsymbol{P}$ and $\boldsymbol{u}'=\boldsymbol{T}\boldsymbol{u}$, $\boldsymbol{P}'=\boldsymbol{k}'\boldsymbol{u}'$,

$$\boldsymbol{P}=\boldsymbol{T}^{-1}\boldsymbol{P}'=\boldsymbol{T}^{-1}\boldsymbol{k}'\boldsymbol{T}\boldsymbol{u}\quad\text{or}\quad \boldsymbol{k}=\boldsymbol{T}^{-1}\boldsymbol{k}'\boldsymbol{T}. \tag{1.43}$$

### 1.5.2 Obliquely Connected Components

Consider a system consisting of components $k_a$ and $k_b$ as is shown in Fig. 1.18. The nodes are numbered as shown in the figure.

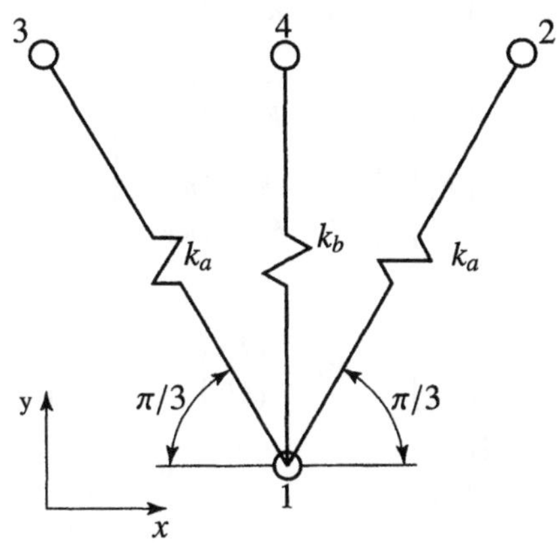


Figure 1.18: Elastic components intersecting obliquely

Now we continue the analysis following the steps explained in subsection 1.3.2.

**(1) Construction of element stiffness** Utilizing the result of the preceding subsection, the element stiffness of each component is obtained from (1.42) as follows:

Component $k_a$, having nodes 1 and 2 at both ends: $c = 1/2$, $s = \sqrt{3}/2$,

$$\begin{Bmatrix} P_1 \\ Q_1 \\ \hline P_2 \\ Q_2 \end{Bmatrix} = \frac{k_a}{4} \left[ \begin{array}{cc|cc} 1 & \sqrt{3} & -1 & -\sqrt{3} \\ \sqrt{3} & 3 & -\sqrt{3} & -3 \\ \hline -1 & -\sqrt{3} & 1 & \sqrt{3} \\ -\sqrt{3} & -3 & \sqrt{3} & 3 \end{array} \right] \begin{Bmatrix} u_1 \\ v_1 \\ \hline u_2 \\ v_2 \end{Bmatrix}. \tag{1.44}$$

Component $k_b$, having nodes 1 and 3 at both ends: $c = 0$, $s = 1$,

$$\begin{Bmatrix} P_1 \\ Q_1 \\ \hline P_3 \\ Q_3 \end{Bmatrix} = k_b \left[ \begin{array}{cc|cc} 0 & 0 & 0 & 0 \\ 0 & 1 & 0 & -1 \\ \hline 0 & 0 & 0 & 0 \\ 0 & -1 & 0 & 1 \end{array} \right] \begin{Bmatrix} u_1 \\ v_1 \\ \hline u_3 \\ v_3 \end{Bmatrix}. \tag{1.45}$$

Component $k_a$, having nodes 1 and 4 at both ends: $c = -1/2$, $s = \sqrt{3}/2$,

$$\begin{Bmatrix} P_1 \\ Q_1 \\ \hline P_4 \\ Q_4 \end{Bmatrix} = \frac{k_a}{4} \left[ \begin{array}{cc|cc} 1 & -\sqrt{3} & -1 & \sqrt{3} \\ -\sqrt{3} & 3 & \sqrt{3} & -3 \\ \hline -1 & \sqrt{3} & 1 & -\sqrt{3} \\ \sqrt{3} & -3 & -\sqrt{3} & 3 \end{array} \right] \begin{Bmatrix} u_1 \\ v_1 \\ \hline u_4 \\ v_4 \end{Bmatrix}. \tag{1.46}$$

Introducing the submatrixes (subdivided matrixes) into the equation to simplify the expression, we get renewed expressions for (1.44) ~ (1.46) as follows;

$$\left. \begin{array}{l} \begin{Bmatrix} \boldsymbol{P}_1 \\ \boldsymbol{P}_2 \end{Bmatrix} = \begin{bmatrix} \boldsymbol{k}_a & -\boldsymbol{k}_a \\ -\boldsymbol{k}_a & \boldsymbol{k}_a \end{bmatrix} \begin{Bmatrix} \boldsymbol{u}_1 \\ \boldsymbol{u}_2 \end{Bmatrix} \\ \begin{Bmatrix} \boldsymbol{P}_1 \\ \boldsymbol{P}_3 \end{Bmatrix} = \begin{bmatrix} \boldsymbol{k}_b & -\boldsymbol{k}_b \\ -\boldsymbol{k}_b & \boldsymbol{k}_b \end{bmatrix} \begin{Bmatrix} \boldsymbol{u}_1 \\ \boldsymbol{u}_3 \end{Bmatrix} \\ \begin{Bmatrix} \boldsymbol{P}_1 \\ \boldsymbol{P}_4 \end{Bmatrix} = \begin{bmatrix} \boldsymbol{k}_c & -\boldsymbol{k}_c \\ -\boldsymbol{k}_c & \boldsymbol{k}_c \end{bmatrix} \begin{Bmatrix} \boldsymbol{u}_1 \\ \boldsymbol{u}_4 \end{Bmatrix} \end{array} \right\}, \tag{1.47}$$

where,

$$\boldsymbol{P}_1 = \begin{Bmatrix} P_1 \\ Q_1 \end{Bmatrix}, \quad \boldsymbol{P}_2 = \begin{Bmatrix} P_2 \\ Q_2 \end{Bmatrix}, \quad \boldsymbol{P}_3 = \begin{Bmatrix} P_3 \\ Q_3 \end{Bmatrix}, \quad \boldsymbol{P}_4 = \begin{Bmatrix} P_4 \\ Q_4 \end{Bmatrix},$$

$$\boldsymbol{u}_1 = \begin{Bmatrix} u_1 \\ v_1 \end{Bmatrix}, \quad \boldsymbol{u}_2 = \begin{Bmatrix} u_2 \\ v_2 \end{Bmatrix}, \quad \boldsymbol{u}_3 = \begin{Bmatrix} u_3 \\ v_3 \end{Bmatrix}, \quad \boldsymbol{u}_4 = \begin{Bmatrix} u_4 \\ v_4 \end{Bmatrix},$$

$$\boldsymbol{k}_a = \frac{k_a}{4} \begin{bmatrix} 1 & \sqrt{3} \\ \sqrt{3} & 1 \end{bmatrix}, \quad \boldsymbol{k}_b = k_b \begin{bmatrix} 0 & 0 \\ 0 & 1 \end{bmatrix}, \quad \boldsymbol{k}_c = \frac{k_a}{4} \begin{bmatrix} 1 & -\sqrt{3} \\ -\sqrt{3} & 3 \end{bmatrix}.$$

**(2) Construction of global stiffness** Using the expression of (1.47), the equilibrium equation of each component can be stated as follows:

$$\begin{Bmatrix} \boldsymbol{P}_1 \\ \boldsymbol{P}_2 \\ \boldsymbol{P}_3 \\ \boldsymbol{P}_4 \end{Bmatrix} = \begin{bmatrix} \boldsymbol{k}_a & -\boldsymbol{k}_a & \boldsymbol{0} & \boldsymbol{0} \\ -\boldsymbol{k}_a & \boldsymbol{k}_a & \boldsymbol{0} & \boldsymbol{0} \\ \boldsymbol{0} & \boldsymbol{0} & \boldsymbol{0} & \boldsymbol{0} \\ \boldsymbol{0} & \boldsymbol{0} & \boldsymbol{0} & \boldsymbol{0} \end{bmatrix} \begin{Bmatrix} \boldsymbol{u}_1 \\ \boldsymbol{u}_2 \\ \boldsymbol{u}_3 \\ \boldsymbol{u}_4 \end{Bmatrix},$$

$$\begin{Bmatrix} \boldsymbol{P}_1 \\ \boldsymbol{P}_2 \\ \boldsymbol{P}_3 \\ \boldsymbol{P}_4 \end{Bmatrix} = \begin{bmatrix} \boldsymbol{k}_b & \boldsymbol{0} & -\boldsymbol{k}_b & \boldsymbol{0} \\ \boldsymbol{0} & \boldsymbol{0} & \boldsymbol{0} & \boldsymbol{0} \\ -\boldsymbol{k}_b & \boldsymbol{0} & \boldsymbol{k}_b & \boldsymbol{0} \\ \boldsymbol{0} & \boldsymbol{0} & \boldsymbol{0} & \boldsymbol{0} \end{bmatrix} \begin{Bmatrix} \boldsymbol{u}_1 \\ \boldsymbol{u}_2 \\ \boldsymbol{u}_3 \\ \boldsymbol{u}_4 \end{Bmatrix},$$

$$\begin{Bmatrix} \boldsymbol{P}_1 \\ \boldsymbol{P}_2 \\ \boldsymbol{P}_3 \\ \boldsymbol{P}_4 \end{Bmatrix} = \begin{bmatrix} \boldsymbol{k}_c & \boldsymbol{0} & \boldsymbol{0} & -\boldsymbol{k}_c \\ \boldsymbol{0} & \boldsymbol{0} & \boldsymbol{0} & \boldsymbol{0} \\ \boldsymbol{0} & \boldsymbol{0} & \boldsymbol{0} & \boldsymbol{0} \\ -\boldsymbol{k}_c & \boldsymbol{0} & \boldsymbol{0} & \boldsymbol{k}_c \end{bmatrix} \begin{Bmatrix} \boldsymbol{u}_1 \\ \boldsymbol{u}_2 \\ \boldsymbol{u}_3 \\ \boldsymbol{u}_4 \end{Bmatrix}.$$

Using the direct stiffness method, the global stiffness can be constructed as follows:

$$\begin{Bmatrix} \boldsymbol{P}_1 \\ \boldsymbol{P}_2 \\ \boldsymbol{P}_3 \\ \boldsymbol{P}_4 \end{Bmatrix} = \begin{bmatrix} \boldsymbol{k}_{11} & \boldsymbol{k}_{12} & \boldsymbol{k}_{13} & \boldsymbol{k}_{14} \\ \boldsymbol{k}_{21} & \boldsymbol{k}_{22} & \boldsymbol{k}_{23} & \boldsymbol{k}_{24} \\ \boldsymbol{k}_{31} & \boldsymbol{k}_{32} & \boldsymbol{k}_{33} & \boldsymbol{k}_{34} \\ \boldsymbol{k}_{41} & \boldsymbol{k}_{42} & \boldsymbol{k}_{43} & \boldsymbol{k}_{44} \end{bmatrix} \begin{Bmatrix} \boldsymbol{u}_1 \\ \boldsymbol{u}_2 \\ \boldsymbol{u}_3 \\ \boldsymbol{u}_4 \end{Bmatrix}, \tag{1.48}$$

where,

$$\boldsymbol{k}_{11} = \boldsymbol{k}_a + \boldsymbol{k}_b + \boldsymbol{k}_c = \begin{bmatrix} \frac{k_a}{2} & 0 \\ 0 & \frac{3k_a}{2} + k_b \end{bmatrix},$$

$$\boldsymbol{k}_{12} = \boldsymbol{k}_{21} = -\boldsymbol{k}_a, \quad \boldsymbol{k}_{13} = \boldsymbol{k}_{31} = -\boldsymbol{k}_b, \quad \boldsymbol{k}_{14} = \boldsymbol{k}_{41} = -\boldsymbol{k}_c,$$

$$\boldsymbol{k}_{22} = \boldsymbol{k}_a, \quad \boldsymbol{k}_{23} = \boldsymbol{k}_{32} = \boldsymbol{0},$$

$$\boldsymbol{k}_{33} = \boldsymbol{k}_b, \quad \boldsymbol{k}_{34} = \boldsymbol{0},$$

$$\boldsymbol{k}_{44} = \boldsymbol{k}_c.$$

Please note that $\boldsymbol{k}_a$, $\boldsymbol{k}_b$ and $\boldsymbol{k}_c$ are submatrixes explained previously. As each component of the coefficient matrix in (1.48) is a matrix of $2\times2$, the size of the total coefficient matrix becomes $8 \times 8$.

**(3) Consideration of boundary conditions** Two cases, A and B, are presented here. In both cases, the nodes 2, 3, and 4 are fixed to an overhead beam.

$$\boldsymbol{u}_2 = \boldsymbol{u}_3 = \boldsymbol{u}_4 = \begin{Bmatrix} 0 \\ 0 \end{Bmatrix}. \tag{1.49}$$

Substituting (1.49) for (1.48), we get

$$\begin{Bmatrix} \boldsymbol{P}_1 \\ \boldsymbol{P}_2 \\ \boldsymbol{P}_4 \\ \boldsymbol{P}_3 \end{Bmatrix} = \begin{bmatrix} \boldsymbol{k}_{11} & \boldsymbol{k}_{12} & \boldsymbol{k}_{13} & \boldsymbol{k}_{14} \\ \boldsymbol{k}_{21} & \boldsymbol{k}_{22} & \boldsymbol{k}_{23} & \boldsymbol{k}_{24} \\ \boldsymbol{k}_{31} & \boldsymbol{k}_{32} & \boldsymbol{k}_{33} & \boldsymbol{k}_{34} \\ \boldsymbol{k}_{41} & \boldsymbol{k}_{42} & \boldsymbol{k}_{43} & \boldsymbol{k}_{44} \end{bmatrix} \begin{Bmatrix} \boldsymbol{u}_1 \\ \boldsymbol{0} \\ \boldsymbol{0} \\ \boldsymbol{0} \end{Bmatrix} = \begin{bmatrix} \boldsymbol{k}_{11} \\ \boldsymbol{k}_{21} \\ \boldsymbol{k}_{31} \\ \boldsymbol{k}_{41} \end{bmatrix} \begin{Bmatrix} u_1 \\ v_1 \end{Bmatrix}. \tag{1.50}$$

[Case A] Suppose the case of Fig. 1.19, where the external force $Q$ is applied in the direction of $-y$.

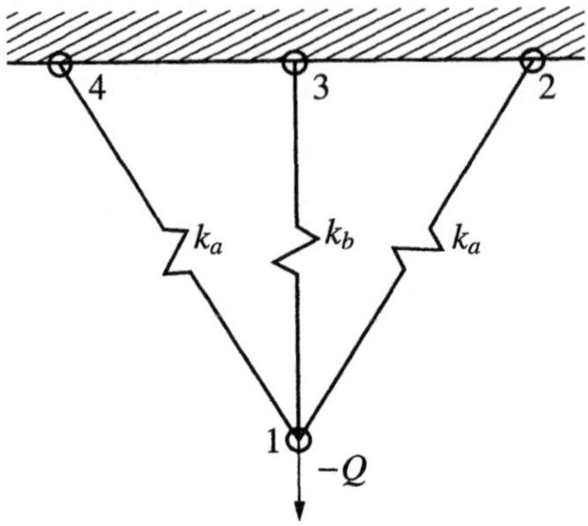


Figure 1.19: A case receiving a vertical force

Substituting the mechanical boundary condition

$$\boldsymbol{P}_1 = \begin{Bmatrix} 0 \\ -Q \end{Bmatrix} \tag{1.51}$$

for (1.50), the first equation can be solved

$$\begin{Bmatrix} u_1 \\ v_1 \end{Bmatrix} = \frac{4}{k_a(3k_a + 2k_b)} \begin{bmatrix} \frac{3k_a}{2} + k_b & \\ 0 & \frac{k_a}{2} \end{bmatrix} \begin{Bmatrix} 0 \\ -Q \end{Bmatrix} = \begin{Bmatrix} 0 \\ \frac{-2Q}{(3k_a + 2k_b)} \end{Bmatrix}. \tag{1.52}$$

The reaction force is obtained by substituting (1.52) for the remaining equations of (1.50)

$$\begin{Bmatrix} P_2 \\ Q_2 \\ P_3 \\ Q_3 \\ P_4 \\ Q_4 \end{Bmatrix} = \frac{Q}{2(3k_a + 2k_b)} \begin{Bmatrix} \sqrt{3}k_a \\ 3k_a \\ 0 \\ 4k_b \\ -\sqrt{3}k_a \\ 3k_a \end{Bmatrix}. \tag{1.53}$$

The condition of static equilibrium shall be fulfilled in the total system.

$$x \text{ direction: } \sum_{i=1}^{4} P_i = \frac{Q}{2(3k_a + 2k_b)}(\sqrt{3}k_a + 0 - \sqrt{3}k_a) = 0,$$

$$y \text{ direction: } \sum_{i=1}^{4} Q_i = -Q + \frac{Q}{2(3k_a + 2k_b)}(3k_a + 4k_b + 3k_a) = 0.$$

[Case B] In the case of Fig. 1.20, where the external force $P$ is applied in the direction of $x$, the mechanical boundary condition is (1.54).

$$\boldsymbol{P}_1 = \begin{Bmatrix} P \\ 0 \end{Bmatrix}. \tag{1.54}$$

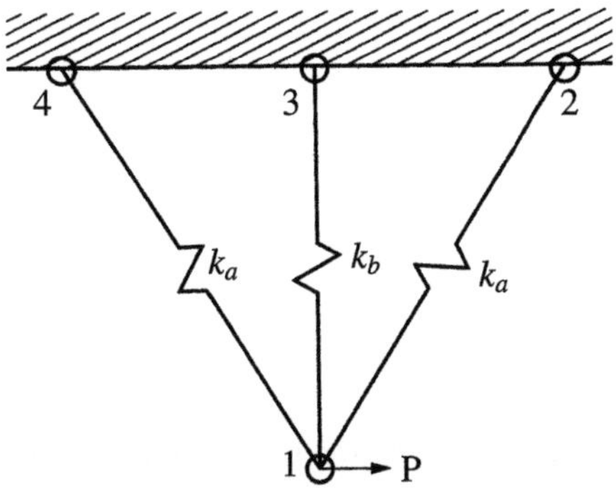


Figure 1.20: A case applying a horizontal force

Substituting (1.54) for (1.50), the first equation will be solved and we get

$$\begin{Bmatrix} u_1 \\ v_1 \end{Bmatrix} = \frac{4}{k_b(3k_a + 2k_b)} \begin{bmatrix} \dfrac{3k_a}{2} + k_b & 0 \\ 0 & \dfrac{k_a}{2} \end{bmatrix} \begin{Bmatrix} P \\ 0 \end{Bmatrix} = \begin{Bmatrix} \dfrac{2P}{k_a} \\ 0 \end{Bmatrix}. \tag{1.55}$$

The reaction force can be obtained by substituting (1.55) for the remaining equations of (1.50)

$$\begin{Bmatrix} P_2 \\ Q_2 \\ P_3 \\ Q_3 \\ P_4 \\ Q_4 \end{Bmatrix} = \frac{P}{2} \begin{Bmatrix} -1 \\ -\sqrt{3} \\ 0 \\ 0 \\ -1 \\ \sqrt{3} \end{Bmatrix}. \tag{1.56}$$

The total system must satisfy the condition of static equilibrium.

$$x \text{ direction:} \sum_{i=1}^{4} P_i = \frac{P}{2}(2 - 1 - 1) = 0,$$

$$y \text{ direction:} \sum_{i=1}^{4} Q_i = \frac{\sqrt{3}}{2} P(1 - 1) = 0.$$

### 1.5.3 From Components to Finite Elements

The contour of the plane domain (a) in Fig. 1.21 is approximated by lines to a polygon (b) and this approximated domain is divided into three elements A, B and C as shown in (c).

Here, the degree of freedom within the plane is assumed to be two ($x$ and $y$ direction). At every vertex of the triangles, the nodes are numbered as is shown in

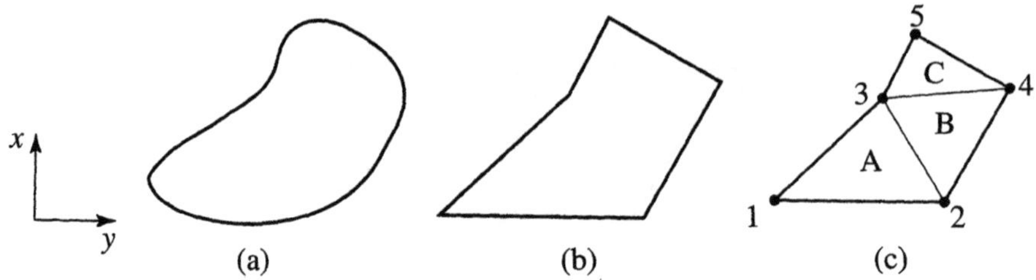


Figure 1.21: Division of a plane domain

Fig. 1.21(c). The correspondence between the element domain and its constituent nodes is tabulated in Table 1.6.

Table 1.6: Correspondence between element domain and nodes

| Element | Constituting nodes |
|---|---|
| A | 1, 2, 3 |
| B | 2, 4, 3 |
| C | 4, 5, 3 |

Element name is defined counterclockwise.

The nodes which are jointly owned by two or more elements are

Node 2: A, B,

Node 3: A, B, C,

Node 4: B, C.

Suppose the elements are in a disjointed condition, for the time being and the stiffness of each element is expressed as the following equations:

$$\begin{Bmatrix} \boldsymbol{P}_1^A \\ \boldsymbol{P}_2^A \\ \boldsymbol{P}_3^A \end{Bmatrix} = \begin{bmatrix} \boldsymbol{k}_{11}^A & \boldsymbol{k}_{12}^A & \boldsymbol{k}_{13}^A \\ \boldsymbol{k}_{21}^A & \boldsymbol{k}_{22}^A & \boldsymbol{k}_{23}^A \\ \boldsymbol{k}_{31}^A & \boldsymbol{k}_{32}^A & \boldsymbol{k}_{33}^A \end{bmatrix} \begin{Bmatrix} \boldsymbol{u}_1^A \\ \boldsymbol{u}_2^A \\ \boldsymbol{u}_3^A \end{Bmatrix}, \tag{1.57}$$

$$\begin{Bmatrix} \boldsymbol{P}_2^B \\ \boldsymbol{P}_4^B \\ \boldsymbol{P}_3^B \end{Bmatrix} = \begin{bmatrix} \boldsymbol{k}_{22}^B & \boldsymbol{k}_{24}^B & \boldsymbol{k}_{23}^B \\ \boldsymbol{k}_{42}^B & \boldsymbol{k}_{44}^B & \boldsymbol{k}_{43}^B \\ \boldsymbol{k}_{32}^B & \boldsymbol{k}_{34}^B & \boldsymbol{k}_{33}^B \end{bmatrix} \begin{Bmatrix} \boldsymbol{u}_2^B \\ \boldsymbol{u}_4^B \\ \boldsymbol{u}_3^B \end{Bmatrix}, \tag{1.58}$$

$$\begin{Bmatrix} \boldsymbol{P}_4^C \\ \boldsymbol{P}_5^C \\ \boldsymbol{P}_3^C \end{Bmatrix} = \begin{bmatrix} \boldsymbol{k}_{44}^C & \boldsymbol{k}_{45}^C & \boldsymbol{k}_{43}^C \\ \boldsymbol{k}_{54}^C & \boldsymbol{k}_{55}^C & \boldsymbol{k}_{53}^C \\ \boldsymbol{k}_{34}^C & \boldsymbol{k}_{35}^C & \boldsymbol{k}_{33}^C \end{bmatrix} \begin{Bmatrix} \boldsymbol{u}_4^C \\ \boldsymbol{u}_5^C \\ \boldsymbol{u}_3^C \end{Bmatrix}. \tag{1.59}$$

The superscript indicates the element name, and the subscript of $\boldsymbol{P}$ and $\boldsymbol{u}$ shows the number of nodes. $\boldsymbol{P}$ and $\boldsymbol{u}$ are vectors on $x$-$y$ plane. By expanding these equations to the total degree of freedom, the above equations can be expressed as follows:

$$\begin{Bmatrix} \boldsymbol{P}_1^A \\ \boldsymbol{P}_2^A \\ \boldsymbol{P}_3^A \\ \boldsymbol{0} \\ \boldsymbol{0} \end{Bmatrix} = \begin{bmatrix} \boldsymbol{k}_{11}^A & \boldsymbol{k}_{12}^A & \boldsymbol{k}_{13}^A & \boldsymbol{0} & \boldsymbol{0} \\ \boldsymbol{k}_{21}^A & \boldsymbol{k}_{22}^A & \boldsymbol{k}_{23}^A & \boldsymbol{0} & \boldsymbol{0} \\ \boldsymbol{k}_{31}^A & \boldsymbol{k}_{32}^A & \boldsymbol{k}_{33}^A & \boldsymbol{0} & \boldsymbol{0} \\ \boldsymbol{0} & \boldsymbol{0} & \boldsymbol{0} & \boldsymbol{0} & \boldsymbol{0} \\ \boldsymbol{0} & \boldsymbol{0} & \boldsymbol{0} & \boldsymbol{0} & \boldsymbol{0} \end{bmatrix} \begin{Bmatrix} \boldsymbol{u}_1 \\ \boldsymbol{u}_2 \\ \boldsymbol{u}_3 \\ \boldsymbol{u}_4 \\ \boldsymbol{u}_5 \end{Bmatrix}, \tag{1.60}$$

$$\begin{Bmatrix} \boldsymbol{0} \\ \boldsymbol{P}_2^B \\ \boldsymbol{P}_3^B \\ \boldsymbol{P}_4^B \\ \boldsymbol{0} \end{Bmatrix} = \begin{bmatrix} \boldsymbol{0} & \boldsymbol{0} & \boldsymbol{0} & \boldsymbol{0} & \boldsymbol{0} \\ \boldsymbol{0} & \boldsymbol{k}_{22}^B & \boldsymbol{k}_{23}^B & \boldsymbol{k}_{24}^B & \boldsymbol{0} \\ \boldsymbol{0} & \boldsymbol{k}_{32}^B & \boldsymbol{k}_{33}^B & \boldsymbol{k}_{34}^B & \boldsymbol{0} \\ \boldsymbol{0} & \boldsymbol{k}_{42}^B & \boldsymbol{k}_{43}^B & \boldsymbol{k}_{44}^B & \boldsymbol{0} \\ \boldsymbol{0} & \boldsymbol{0} & \boldsymbol{0} & \boldsymbol{0} & \boldsymbol{0} \end{bmatrix} \begin{Bmatrix} \boldsymbol{u}_1 \\ \boldsymbol{u}_2 \\ \boldsymbol{u}_3 \\ \boldsymbol{u}_4 \\ \boldsymbol{u}_5 \end{Bmatrix}, \tag{1.61}$$

$$\begin{Bmatrix} \boldsymbol{0} \\ \boldsymbol{0} \\ \boldsymbol{P}_3^C \\ \boldsymbol{P}_4^C \\ \boldsymbol{P}_5^C \end{Bmatrix} = \begin{bmatrix} \boldsymbol{0} & \boldsymbol{0} & \boldsymbol{0} & \boldsymbol{0} & \boldsymbol{0} \\ \boldsymbol{0} & \boldsymbol{0} & \boldsymbol{0} & \boldsymbol{0} & \boldsymbol{0} \\ \boldsymbol{0} & \boldsymbol{0} & \boldsymbol{k}_{33}^C & \boldsymbol{k}_{34}^C & \boldsymbol{k}_{35}^C \\ \boldsymbol{0} & \boldsymbol{0} & \boldsymbol{k}_{43}^C & \boldsymbol{k}_{44}^C & \boldsymbol{k}_{45}^C \\ \boldsymbol{0} & \boldsymbol{0} & \boldsymbol{k}_{53}^C & \boldsymbol{k}_{54}^C & \boldsymbol{k}_{55}^C \end{bmatrix} \begin{Bmatrix} \boldsymbol{u}_1 \\ \boldsymbol{u}_2 \\ \boldsymbol{u}_3 \\ \boldsymbol{u}_4 \\ \boldsymbol{u}_5 \end{Bmatrix}. \tag{1.62}$$

The whole system can be expressed in equation (1.63),

$$\begin{Bmatrix} \boldsymbol{P}_1 \\ \boldsymbol{P}_2 \\ \boldsymbol{P}_3 \\ \boldsymbol{P}_4 \\ \boldsymbol{P}_5 \end{Bmatrix} = \begin{bmatrix} \boldsymbol{k}_{11}^A & \boldsymbol{k}_{12}^A & \boldsymbol{k}_{13}^A & \boldsymbol{0} & \boldsymbol{0} \\ \boldsymbol{k}_{21}^A & (\boldsymbol{k}_{22}^A + \boldsymbol{k}_{22}^B) & (\boldsymbol{k}_{23}^A + \boldsymbol{k}_{23}^B) & \boldsymbol{k}_{24}^B & \boldsymbol{0} \\ \boldsymbol{k}_{31}^A & (\boldsymbol{k}_{32}^A + \boldsymbol{k}_{32}^B) & (\boldsymbol{k}_{33}^A + \boldsymbol{k}_{33}^B + \boldsymbol{k}_{33}^C) & (\boldsymbol{k}_{34}^B + \boldsymbol{k}_{34}^C) & \boldsymbol{k}_{35}^C \\ \boldsymbol{0} & \boldsymbol{k}_{42}^B & (\boldsymbol{k}_{43}^B + \boldsymbol{k}_{43}^C) & (\boldsymbol{k}_{44}^B + \boldsymbol{k}_{44}^C) & \boldsymbol{k}_{45}^C \\ \boldsymbol{0} & \boldsymbol{0} & \boldsymbol{k}_{53}^C & \boldsymbol{k}_{54}^C & \boldsymbol{k}_{55}^C \end{bmatrix} \begin{Bmatrix} \boldsymbol{u}_1 \\ \boldsymbol{u}_2 \\ \boldsymbol{u}_3 \\ \boldsymbol{u}_4 \\ \boldsymbol{u}_5 \end{Bmatrix} \tag{1.63}$$

where,

$$\boldsymbol{P}_1 = \boldsymbol{P}_1^A, \quad \boldsymbol{P}_2 = \boldsymbol{P}_1^A + \boldsymbol{P}_2^B, \quad \boldsymbol{P}_3 = \boldsymbol{P}_3^A + \boldsymbol{P}_3^B + \boldsymbol{P}_3^C, \quad \boldsymbol{P}_4 = \boldsymbol{P}_4^B + \boldsymbol{P}_4^C, \quad \boldsymbol{P}_5 = \boldsymbol{P}_5^C.$$

The coefficient matrix is a global stiffness matrix. The global matrix is symmetric, since $\boldsymbol{k}_{ij}$ constituting each component are symmetric. In the coefficient matrix, the component is equivalent to the sum of the stiffness of the elements to which the node in concern belongs commonly.

Next we give the boundary condition. Let us hypothesize a case, for example, where the nodes 1 and 2 are fixed and an external force $P$ is applied at node 5.

$$\text{Geometrical boundary condition: } \boldsymbol{u}_1 = \boldsymbol{u}_2 = \boldsymbol{0},$$

$$\text{Mechanical boundary condition: } \boldsymbol{P}_3 = \boldsymbol{P}_4 = \boldsymbol{0}, \quad \boldsymbol{P}_5 = \boldsymbol{P}.$$

Our next goal is to solve the six-dimensional simultaneous linear equation corresponding from third to fifth row in (1.63) with respect to the unknown displacement $\boldsymbol{u}$, considering the above boundary conditions.

$$\begin{Bmatrix} \boldsymbol{0} \\ \boldsymbol{0} \\ \boldsymbol{P} \end{Bmatrix} = \begin{bmatrix} (\boldsymbol{k}_{33}^A + \boldsymbol{k}_{33}^B + \boldsymbol{k}_{33}^C) & (\boldsymbol{k}_{34}^B + \boldsymbol{k}_{34}^C) & \boldsymbol{k}_{35}^C \\ (\boldsymbol{k}_{43}^B + \boldsymbol{k}_{43}^C) & (\boldsymbol{k}_{44}^B + \boldsymbol{k}_{44}^C) & \boldsymbol{k}_{45}^C \\ \boldsymbol{k}_{53}^C & \boldsymbol{k}_{54}^C & \boldsymbol{k}_{55}^C \end{bmatrix} \begin{Bmatrix} \boldsymbol{u}_3 \\ \boldsymbol{u}_4 \\ \boldsymbol{u}_5 \end{Bmatrix}. \tag{1.64}$$

In the present discussion, to make the flow of analysis easier to understand, the approximation was carried out by dividing the domain into only three parts. When the domain is divided into a larger number of triangles, the accuracy will improve. The reader should understand that the two-dimensional problem can be treated with the same step used for an elastic component. It is easy to imagine, based on the aforementioned discussion, that the three-dimensional problem is also treated in the same way. An increase of the degree of freedom simply means there is an increase of the number of dimensions in the simultaneous equation. The only knowledge which is to be acquired in solving the problem of three dimensional construction is a method to derive an element stiffness.

## Bibliography of Chapter 1

[1] Wittrick W H, Int. J. of Mechanical Science (1965), p25
[2] Togawa H, Numerical Calculation of Matrix (in Japanese), OHM SHA (1971)

# 2 OUTLINE OF FINITE ELEMENT METHOD (FEM)

The derivation of element stiffness will be discussed in this chapter. When element stiffness matrix is given, the solution will be obtained through the following process:

1. The construction of global stiffness matrix through the assembling of given element stiffness matrixes

2. The provision of the boundary condition

3. The solution of the simultaneous equation

This process was explained in paragraph 1.5.3. The derivation of element stiffness requires a fundamental knowledge of the material dynamics. To meet with this requirement, some fundamental items relating to elastic and plastic mechanics are provided in section 2.1.

Since FEM is a direct numerical solution, the reader only needs to comprehend the meanings of some technical terms.

In section 2.2, the nonlinearity problem indispensable to the analysis in chapter 6 will be introduced. In section 2.3, the treatment of time-dependent dynamic problems, such as vibration or impact, will be discussed.

In the last section 2.4, the concrete derivating process of element stiffness will be provided, based on some examples of truss elements, plane elements, and three-dimensional elastic bodies.

## 2.1 FUNDAMENTALS OF ELASTO-PLASTICITY DYNAMICS

### 2.1.1 Viewpoint of Continuum Dynamics

The continuumis a general term used for elastic bodies, fluid, viscoelastic bodies, and so on.

In continuum dynamics, a common perspective is required irrespective of the material or kind of the object. It recognizes the object with two viewpoints. One is dynamic and the other is geometrical viewpoint.

These can be divided into two groups:

- Dynamic quantities: stress, body force, surface force

- Geometrical quantities: strain, displacement (strain rate)

The entire space occupies the inside and the outside of the domain. As a continuum exists in the domain, general equations over the field of the domain should be fundamentally ones for the continuum. The continuum borders the outside of the domain through a boundary, as long as the continuum has a limited domain. It naturally follows that any involvement of the continuum with the outside of the domain should occur on the boundary. This restraining condition is called the boundary condition. Since the continuum is described by the mechanical quantity and the geometrical quantity, there is a mechanical boundary condition and a geometrical boundary condition.

### 2.1.2 General Equations

The general equation concerning the continuum consists of the following three equations:

- Equilibrium equation
- Displacement-strain equation
- Constitutive equation

The equilibrium equation, is the relationship between stress and body force and describes the static balance of the mechanical quantities of minute element. The displacement-strain equation indicates the geometrical continuity and ensures that no crack or overlapping occurs from deformation. While these two equations apply to any continuum, the constitutive equation specifies the characteristic of each continuum in concern. In other words, this equation plays a role in joining the mechanical and the geometrical world and is basically given as a relationship between stress and strain.

The number of variables shall be equal to the number of equations. This is the prerequisite for any problem solving. In the case of a three-dimensional linear elastic body, counting stands as follows:

| | | | |
|---|---|---|---|
| Independent components of stress | = 6 | Equilibrium equations | = 3 |
| Independent components of strain | = 6 | Displacement-strain equations | = 6 |
| Components of displacement | = 3 | Hooke's laws | = 6 |
| Variables | = 15 | Equations | = 15 |

Details of the equation will be described in the next paragraph.

There are two methods to solve these 15 equations. One is displacement method and the other is stress method. The selection of the method depends on which quantity is to be treated as the variables. FEM also has two methods corresponding to the variables mentioned above. The displacement method will be introduced in this book.

### 2.1.3 Basic Equations for Linear Elastic Body

**(1) Tensor notation** Suppose a continuum is placed in a three-dimensional space with an orthogonal coordinate consisting of $x_1$, $x_2$, and $x_3$. The displacement and the body force can be expressed as (2.1), since they are vectors having components along each direction.

$$u_i = \begin{Bmatrix} u_1 \\ u_2 \\ u_3 \end{Bmatrix}, \quad f_i = \begin{Bmatrix} f_1 \\ f_2 \\ f_3 \end{Bmatrix}. \tag{2.1}$$

Stress and strain are second-order tensors and can be expressed as (2.2).

$$\sigma_{ij} = \begin{bmatrix} \sigma_{11} & \sigma_{12} & \sigma_{13} \\ & \sigma_{22} & \sigma_{23} \\ \text{SYM.} & & \sigma_{33} \end{bmatrix}, \quad \varepsilon_{ij} = \begin{bmatrix} \varepsilon_{11} & \varepsilon_{12} & \varepsilon_{13} \\ & \varepsilon_{22} & \varepsilon_{23} \\ \text{SYM.} & & \varepsilon_{33} \end{bmatrix}. \tag{2.2}$$

Since, $ij$ component = $ji$ component, for shear components, both the stress and the strain become the symmetric matrixes. The notation SYM. in (2.2) means a symmetric matrix.

The term of tensor refers to the amount having the magnitude and direction. The order of tensor is based on the number of directions. That is,

| | | |
|---|---|---|
| zero-order tensor | : no direction | = scalar |
| first-order tensor | : one direction | = vector |
| second-order tensor | : two directions | |
| third-order tensor | : three directions | |
| fourth-order tensor | : four directions | |
| ⋮ | ⋮ | |

For example, a stress tensor has two directions, the direction of the virtual surface on which the stress is applied and the direction of force. Therefore, it is a second-order tensor. The magnitude of the tensor component cannot be determined unless the number of directions corresponding to the number of orders is specified. This means that the tensor notation needs as many subscripts as the orders.

The basic equation of a linear elastic body can be expressed as follows, using the above stated notations,

$$\text{Equilibrium equation: } \sigma_{ij,j} + f_i = 0, \tag{2.3}$$

$$\text{Displacement-strain equation: } \varepsilon_{ij} = \frac{1}{2}(u_{i,j} + u_{j,i}), \tag{2.4}$$

$$\text{Stress-strain equation: } \sigma_{ij} = D_{ijkl}\varepsilon_{kl}. \tag{2.5}$$

The fourth-order tensor $D_{ijkl}$ is the elastic coefficient and has 36 ( $= 6 \times 6$) components, even if the symmetry of the stress and the strain is taken into account. We should remember that introducing the assumption of isotropy reduces the number of independent elastic coefficient components to two ($E$ and $G$ shown below).

For this we use,

$E$: Young's modulus, $G$: Rigidity modulus, $\nu$: Poisson's ratio,

$$G = \frac{E}{2}(1+\nu).$$

**(2) Vector notation** The notation used above is compact and good for treating mathematical formulas, but is not suitable to carry out any practical computation. For instance, $D_{ijkl}$ is a fourth-order tensor and has 81 ( $= 3^4$) original components. However, the sufficient number of components is 36, as stated before, and even decreases to 21, when the reciprocity theorem is taken into account. FEM redefines the six components of the stress and the strain as a one-dimensional vector arrangement as follows,

$$\boldsymbol{\sigma} = \begin{Bmatrix} \sigma_{11} \\ \sigma_{22} \\ \sigma_{33} \\ \sigma_{12} \\ \sigma_{23} \\ \sigma_{31} \end{Bmatrix} = \begin{Bmatrix} \sigma_x \\ \sigma_y \\ \sigma_z \\ \tau_{xy} \\ \tau_{yz} \\ \tau_{zx} \end{Bmatrix}, \quad \boldsymbol{\varepsilon} = \begin{Bmatrix} \varepsilon_{11} \\ \varepsilon_{22} \\ \varepsilon_{33} \\ 2\varepsilon_{12} \\ 2\varepsilon_{23} \\ 2\varepsilon_{31} \end{Bmatrix} = \begin{Bmatrix} \varepsilon_x \\ \varepsilon_y \\ \varepsilon_z \\ \gamma_{xy} \\ \gamma_{yz} \\ \gamma_{zx} \end{Bmatrix}. \tag{2.6}$$

In equation (2.6), the coordinate system is changed from 1, 2, 3 to $x$, $y$, $z$ and the definition of strain is changed to mechanical strain.

By adopting the above notations, the fundamental equations are restated as follows,

$$\text{Equilibrium equation:} \begin{Bmatrix} \dfrac{\partial \sigma_x}{\partial x} + \dfrac{\partial \tau_{xy}}{\partial y} + \dfrac{\partial \tau_{xz}}{\partial z} \\ \dfrac{\partial \tau_{xy}}{\partial x} + \dfrac{\partial \sigma_y}{\partial y} + \dfrac{\partial \tau_{xz}}{\partial z} \\ \dfrac{\partial \tau_{xz}}{\partial x} + \dfrac{\partial \tau_{yz}}{\partial y} + \dfrac{\partial \sigma_z}{\partial z} \end{Bmatrix} + \begin{Bmatrix} f_x \\ f_y \\ f_z \end{Bmatrix} = \begin{Bmatrix} 0 \\ 0 \\ 0 \end{Bmatrix}, \tag{2.7}$$

$$\text{Displacement-strain equation:} \begin{Bmatrix} \varepsilon_x \\ \varepsilon_y \\ \varepsilon_z \\ \gamma_{xy} \\ \gamma_{yz} \\ \gamma_{zx} \end{Bmatrix} = \begin{bmatrix} \partial/\partial x & 0 & 0 \\ 0 & \partial/\partial y & 0 \\ 0 & 0 & \partial/\partial z \\ \partial/\partial y & \partial/\partial x & 0 \\ 0 & \partial/\partial z & \partial/\partial y \\ \partial/\partial z & 0 & \partial/\partial x \end{bmatrix} \begin{Bmatrix} u_x \\ u_y \\ u_z \end{Bmatrix}, \tag{2.8}$$

Stress-strain equation:

$$\begin{Bmatrix} \sigma_x \\ \sigma_y \\ \sigma_z \\ \tau_{xy} \\ \tau_{yz} \\ \tau_{zx} \end{Bmatrix} = \frac{E(1-\nu)}{(1+\nu)(1-2\nu)} \begin{bmatrix} 1 & c_1 & c_1 & 0 & 0 & 0 \\ & 1 & c_1 & 0 & 0 & 0 \\ & & 1 & 0 & 0 & 0 \\ & & & c_2 & 0 & 0 \\ & & & & c_2 & 0 \\ & \text{SYM.} & & & & c_2 \end{bmatrix} \begin{Bmatrix} \varepsilon_x \\ \varepsilon_y \\ \varepsilon_z \\ \gamma_{xy} \\ \gamma_{yz} \\ \gamma_{zx} \end{Bmatrix}, \tag{2.9}$$

where,

$$c_1 = \frac{\nu}{(1-\nu)}, \quad c_2 = \frac{(1-2\nu)}{2(1-\nu)}. \tag{2.10}$$

### 2.1.4 Principle of Virtual Work

Although it is a fundamental rule to solve the equations (2.7), ..., (2.9) on the geometrical and mechanical boundary conditions to obtain the stress or the displacement, the principle of virtual work is often used instead of the equilibrium equation. In this case, three equations — the principle of virtual work, the displacement-strain equation, and the stress-strain equation are to be solved along the geometrical boundary condition. The mechanical boundary condition is included in the principle of virtual work.

The principle of virtual work in particle mechanics can be described as follows:

> "When a particle is in a state of equilibrium, even if some virtual displacement is given to the particle, the work done by the applied force through this displacement is zero."

Using the following notations,

$$\text{Surface force: } \boldsymbol{P} = \begin{Bmatrix} P_x \\ P_y \\ P_z \end{Bmatrix}, \quad \text{Virtual displacement: } \delta\boldsymbol{u} = \begin{Bmatrix} \delta u_x \\ \delta u_y \\ \delta u_z \end{Bmatrix}, \tag{2.11}$$

$$\text{Virtual strain : } \delta\boldsymbol{\varepsilon} = \begin{Bmatrix} \delta\varepsilon_x \\ \delta\varepsilon_y \\ \delta\varepsilon_z \\ \delta\gamma_{xy} \\ \delta\gamma_{yz} \\ \delta\gamma_{zx} \end{Bmatrix} = \begin{bmatrix} \partial/\partial x & 0 & 0 \\ 0 & \partial/\partial y & 0 \\ 0 & 0 & \partial/\partial z \\ \partial/\partial y & \partial/\partial x & 0 \\ 0 & \partial/\partial z & \partial/\partial y \\ \partial/\partial z & 0 & \partial/\partial x \end{bmatrix} \begin{Bmatrix} \delta u_x \\ \delta u_y \\ \delta u_z \end{Bmatrix}. \tag{2.12}$$

The principle can be extended to the case of continuum, obtaining

$$\int_{\Gamma_\sigma} \delta\boldsymbol{u}^T \boldsymbol{P}\, d\Gamma + \int_\Omega \delta\boldsymbol{u}^T \boldsymbol{f}\, d\Omega - \int_\Omega \delta\boldsymbol{\varepsilon}^T \boldsymbol{\sigma}\, d\Omega = 0. \tag{2.13}$$

Here, $\Gamma_\sigma$ express the mechanical boundary.

The physical meaning of (2.13) can be understood in the following manner.

By transposing the third term of the left side of the equation to the right side, we get:

Exterior virtual work = Interior virtual work.

Further, the left side of this equation can be interpreted as:

Exterior virtual work = Virtual work done by surface force
+ Virtual work done by body force.

Please note that stress and strain in the interior virtual work has no relationship to each other. For instance, an equation like (2.9) is not included in the expression. Therefore, the equation (2.13) stands even in the plastic deformation zone. Plastic analysis will be discussed in 2.2.2. As the principle of virtual work is valid in any arbitrary domain of elastic body, the principle is effective to single finite element as well as to the whole domain.

## 2.2 EXPANSION TO NONLINEAR PROBLEMS

### 2.2.1 Geometrical Nonlinearity

The basic method to treat a problem of large elastic deformation is briefly introduced below.

In the basic equations (2.3) ~ (2.5) for linear elastic body, the equation concerning the geometrical quantity was the displacement/strain equation of (2.4). In this equation, the displacement was assumed to be quite small and the terms of the second order or higher were ignored. When the higher order terms are not neglected, nonlinear terms would be added to the equation as follows

$$e_{ij} = \frac{1}{2}(u_{i,j} + u_{j,i} + u_{k,i}u_{k,j}) = \varepsilon_{ij} + \frac{1}{2}(u_{k,i}u_{k,j}). \tag{2.14}$$

In other words, it is necessary to use (2.14) instead of (2.4) to treat the geometrical nonlinearity.

Introducing such strains, the remaining two basic equations are as follows:

$$\text{Equilibrium equation: } [\sigma_{ij}(\delta_{kj} + u_{k,j})]_{,i} + f_k = 0, \tag{2.15}$$

$$\text{Stress-strain equation: } \sigma_{ij} = D_{ijkl}e_{kl}. \tag{2.16}$$

The stress in the above equation is called the Viola-Kirchhoff stress and is defined by the coordinate before deformation.

It should be noted that, in geometrical nonlinear problem, the definition of all the quantities will change depending on whether the coordinate system is fixed to the pre-deformation condition or post-deformation condition.

In our present discussion, Lagrange's method is adopted; this is the method used to describe the movement of a specific point within a fixed space coordinate. So, the stress and the body force in (2.15) are defined prior to the deformation. Also, the mechanical boundary condition is specified on the boundary before the deformation. The principle of virtual work, using Lagrange's method, can be expressed in the same form as (2.13) as follows:

$$\int_{\Gamma_\sigma} \delta \boldsymbol{u}^T \boldsymbol{P}\, d\Gamma + \int_\Omega \delta \boldsymbol{u}^T \boldsymbol{f}\, d\Omega - \int_\Omega \delta \boldsymbol{e}^T \boldsymbol{\sigma}\, d\Omega = 0. \tag{2.17}$$

In the Lagrange method of notation, there are two ways of the formulation for the incremental process. Let's consider an elastic body in initial state O, in which both stress and strain are zero at $t = 0$. When force is applied, the body deforms to state I at time $t$ and further to state II at time $(t + \Delta t)$. If all the quantities are known at state I, there are two methods to determine an increment from state I to state II, depending on the selection of coordinates:

| | | |
|---|---|---|
| Total Lagrange's method | : | to take state O as the origin |
| Update Lagrange's method | : | to take state I as the origin |

The representation of the increment in (2.17) will follow the selected method.

**(1) Residues** The incremental method is a kind of solution in which the nonlinear equation is linearized and connected in small pieces employing tangential gradients at both ends. When using this incremental method, it is necessary to note that each increment does not strictly fit to the original equation at each node (State I in the above example).

To make our discussion more concrete, let us apply the incremental method to (1.1). The differential of (1.1) was (1.2). Let $u_{n+1} = u_n + \Delta u$ and $y_{n+1} = y_n + \Delta y$. Then, the corresponding difference equation can be expressed as

$$\Delta y = y' \Delta u.$$

$y_n$ is an approximate value and is not $y(u_n)$ itself. When

$$R_n = y(u_n) - y_n \neq 0 \tag{2.18}$$

is defined, then,

$$y_n + \Delta y = y(u_n) + y'(u_n)\Delta u + \cdots \tag{2.19}$$

can be expressed as follows by the linear approximation,

$$\Delta y = y'(u_n)\Delta u + R_n. \tag{2.20}$$

The $R_n$ in the above equation is called the residue. This residue plays an important role both in total Lagrange's method and updated Lagrange's method as long as the incremental method is adopted.

**(2) Expression of increment** The equilibrium equation at each increment is expressed as

$$\boldsymbol{K}_T \Delta \boldsymbol{u} = \Delta \boldsymbol{P} - \Delta \boldsymbol{R}, \tag{2.21}$$

where, $\boldsymbol{K}_T$: Tangent stiffness matrix, $\Delta \boldsymbol{u}$: Increment of nodal displacement, $\Delta \boldsymbol{P}$: Increment of nodal force, $\Delta \boldsymbol{R}$: Residue vector. The expression of the tangent stiffness matrix $\boldsymbol{K}_T$ varies depending on which of Lagrange's methods is used. In total Lagrange's method, it is expressed as

$$\boldsymbol{K}_T = \boldsymbol{K}_0 + \boldsymbol{K}_1 + \boldsymbol{K}_2, \tag{2.22}$$

where, $\boldsymbol{K}_0$: Linear stiffness matrix, $\boldsymbol{K}_1$: Primary displacement matrix, $\boldsymbol{K}_2$: Primary stress matrix.

In update Lagrange's method, it is expressed as

$$K_T = K + K_G, \tag{2.23}$$

where, $\boldsymbol{K}$: Linear stiffness matrix, $\boldsymbol{K}_G$: Geometrical stiffness matrix.

In this case, both $\boldsymbol{K}$ and $\boldsymbol{K}_G$ must be calculated at each increment because the coordinate is updated at each increment.

**(3) Problem of contact** The nonlinear problem of springs often involves the problem of contact. The helical spring and the laminated spring, for instance, are sometimes designed to increase the contact area in order to obtain nonlinear properties. We cannot help but consider changes in the contact state in analyzing such phenomenon. A detailed example will be introduced in paragraph 3.4.2.

The beginning of contact during the deformation means that the boundary condition at that stage of increment has been changed from mechanical to geometrical boundary condition:

| | | | |
|---|---|---|---|
| Non-contact | : | Zero external force | (mechanical boundary condition) |
| Contact | : | Constrained displacement | (geometrical boundary condition) |

The separation from contact is a phenomenon opposite to the above. It can be recognized as the changeover from geometrical to mechanical boundary condition.

To analyze the contact problem, it is necessary to proceed to the next step only after reexamining the boundary condition at every increment. The nonlinearity accompanying the amendment of such a boundary condition is sometimes called, 'boundary nonlinearity.' It shall be noted, however, that reexamination of the boundary condition at each step is processed on the assumption of geometrical nonlinearity. It is basically a kind of geometrical nonlinearity.

There are two methods to change the boundary condition. The first one is a method to return, on each increment, to the point in which the boundary condition is incorporated into the global stiffness; the second one is a method to incorporate contact elements into the global stiffness.

In the latter method, the contact elements are inserted in the place (between elements or nodes) where contact is anticipated and the judgment of the contact shall be recognized automatically by prescribed condition. In the general purpose program, it is a main stream to use the contact element. Every software firm has its own sales points to offer better solutions based on the contact element method.

### 2.2.2 Material Nonlinearity

The geometrical nonlinearity where the displacement/strain equation is nonlinear was discussed in the preceding paragraph. In this paragraph, the material nonlinearity where the stress/strain equation is nonlinear will be introduced. The case in which the material nonlinearity shall be handled often accompanies geometrical nonlinearity together and the two nonlinearities are in a close relationship. However, the discussion is restricted here within the nonlinearity of constitutive equation for each case.

In the category of material nonlinearity, elasto-plasticity and the creep are major issues. In this book, no creep is scheduled to be discussed and our theme is limited to the problem of elasto-plasticity. Also, the temperature dependency will not be discussed.

There are two methods in elasto-plastic analysis. These are the strain incremental theory (flow theory) and the total strain theory (deformation theory). The latter is more convenient in terms of the treatment of analysis, but the former is superior in terms of analyzing accuracy. In FEM, the strain incremental theory is more commonly used. Here is an outline of strain incremental theory.

As was previously explained, the material nonlinearity problem can be treated by replacing only the stress/strain equation among the fundamental equations of the linear elastic body. This constitutive equation is expressed in an incremental form, and as a result, other equations shall be unified into incremental expressions:

$$\text{Equilibrium equation: } \sigma_{ij,j} + \Delta f_i = 0, \tag{2.24}$$

$$\text{Displacement/strain equation: } \Delta\varepsilon_{ij} = \frac{1}{2}(u_{i,j} + \Delta u_{j,i}), \tag{2.25}$$

$$\text{Stress/strain equation: } \Delta\sigma_{ij} = D_{ijkl}\Delta\varepsilon_{kl}. \tag{2.26}$$

$D_{ijkl}$ in (2.26) is, of course, the coefficient of elasto-plastic material and its content differs from that of (2.5). Also, the principle of virtual work in (2.13) holds true if the concerned quantities are regarded the increment.

The remaining problem is the content of $D_{ijkl}$. By denoting the increment of the stress and the strain by $\Delta\boldsymbol{\sigma}$ and $\Delta\boldsymbol{\varepsilon}$, respectively, (2.26) is expressed as

$$\Delta\boldsymbol{\sigma} = \boldsymbol{D}\Delta\boldsymbol{\varepsilon}, \quad \boldsymbol{D} = \boldsymbol{D}^e - \boldsymbol{D}^p, \tag{2.27}$$

here, $\boldsymbol{D}^e$ is the elastic coefficient matrix shown in (2.9) and $\boldsymbol{D}^p$ is a coefficient which should be taken into account when the material is within a plastic zone.

The application of the associated stream law, gives us,

$$\boldsymbol{D}^P = \frac{\boldsymbol{D}^e \left\{\frac{\partial f}{\partial \sigma}\right\}\left\{\frac{\partial f}{\partial \sigma}\right\}\boldsymbol{D}^e}{-\left\{\frac{\partial f}{\partial \varepsilon^P}\right\}^T\left\{\frac{\partial f}{\partial \sigma}\right\} + \left\{\frac{\partial f}{\partial \sigma}\right\}^T \boldsymbol{D}^e \left\{\frac{\partial f}{\partial \sigma}\right\}} \tag{2.28}$$

To understand the associated stream law, a background in plastic mechanics is necessary. For more detail, please refer to [1].

The starting condition of plastic deformation beyond the elastic domain is determined by the combination of stresses. It is called a yield condition. The yield condition is generally expressed by a scalar function called the yield function, and is as follows;

$$f(J_1, J_2, J_3) = 0,$$

where $J_1$, $J_2$, and $J_3$ are the stress invariants from the first to third order, respectively. The most famous yield condition for metallic material is Mises's. In this case, $f$ is given as

$$f = \sqrt{3J_2'} - \sigma_0, \tag{2.29}$$

where $J_2'$ is given by (2.30),

$$J_2' = -(\sigma_x'\sigma_y' + \sigma_y'\sigma_z' + \sigma_z'\sigma_x') + \tau_{xy}^2 + \tau_{yz}^2 + \tau_{zx}^2. \tag{2.30}$$

In (2.30), each variable relates to each other as in (2.31).

$$\sigma_x' = \sigma_x - \sigma_m, \quad \sigma_y' = \sigma_y - \sigma_m, \quad \sigma_z' = \sigma_z - \sigma_m, \quad \sigma_m = \frac{\sigma_x + \sigma_y + \sigma_z}{3}. \tag{2.31}$$

The notation $\sigma_0$ in (2.29) refers to the yield stress for the uniaxial tension. This yield condition forms a curved surface in the stress space (a space in which the principal stresses are taken as the coordinates). This curved surface is called the yield surface.

Beyond the yielding, the plastic deformation proceeds while causing strain hardening. In this phase, there is no one-to-one correspondence between the stress and the strain as in the case of the elastic domain. In other words, the strain at a certain time depends both on the history of deformation, as well as the stress conditions at that time.

A rule that specifies the change of the yield surface with the progress of the plastic deformation is called the work hardening rule. The isotropic hardening rule and the moving hardening rule are representative of work hardening rules. In the former instance, the yield surface expands with the hardening, while in the latter, the center moves along with the hardening without changing the size of the yield surface. In both cases, the yield function (2.29) is given as a function of the plastic strain and the increment of the plastic strain is given by the flow rule. The flow rule that uses the plastic potential as the yield function f, is called the associated flow rule. Equation (2.28) derives from this assumption.

## 2.3 EXPANSION TO DYNAMIC PROBLEMS

### 2.3.1 Mass and Damping Matrix

The linear vibration of one degree of freedom can be expressed by (1.6). Extending this to a system of multiple degrees of freedom, gives us

$$\boldsymbol{M}\ddot{\boldsymbol{u}} + \boldsymbol{C}\dot{\boldsymbol{u}} + \boldsymbol{K}\boldsymbol{u} = \boldsymbol{P}, \tag{2.32}$$

where $\boldsymbol{M}$: Mass matrix, $\boldsymbol{C}$: Damping matrix. The special case where $\boldsymbol{C} = \boldsymbol{P} = \boldsymbol{0}$ corresponds to (1.10), and this case is called non-damping free vibration.

In FEM, the extent of space is treated by discretization and the continuum results in a system of multiple degrees of freedom. For that, continuous quantities such as mass or damping force, which are indispensable to the vibration, must be discretized beforehand. Since the basic technique of solution was explained in 1.2.2, a simple idea will be presented here.

**(1) Mass matrix** There are two kinds of mass matrixes based on differences in the construction process. These are the concentrated mass matrix and the consistent mass matrix.

In the concentrated mass matrix, the mass is concentrated at the node and the matrix takes the form of a diagonal matrix such as shown in (1.9). As a result of this, the matrix can be treated as a one-dimensional arrangement and it is relatively simple to obtain an inverse matrix. Although the accuracy of concentrated mass matrix is inferior to the consistent mass matrix, it is often used because of its convenience.

The consistent mass matrix, on the other hand, is derived by converting the mass to the nodal external force equivalent to the inertia force. In this case, there exists a non-zero component in the non-diagonal component in the matrix and this renders the treatment more difficult than the concentrated mass matrix.

The option to choose the kind of mass matrix depends on a situation. The consistent mass matrix is theoretically more orthodox than the concentrated mass matrix.

**(2) Damping matrix** There are many kinds of damping forces. Friction, for example, is a kind of damping force. But the nonlinear damping like friction is complicated to handle. For the present discussion, we will limit ourselves to damping which is proportional to velocity. First, let us recognize the proportional damping consists of two parts[2];

Proportional damping = Viscous damping + Structural damping

Viscous damping is caused by the relative motion of the system against the exterior domain. The damping effect is dependent on the viscosity of fluid occupying the environment. In relation, structural damping is caused by interaction within the system.

In paragraph 1.2.2, the damping term was neglected but the vibration of the chord of a guitar should diminish due to the air resistance at the end in reality. The air resistance in this case corresponds to the viscous damping. Even if a guitar string is plucked in vacuum, its chord will cease to vibrate eventually. In this case, dynamic energy is transformed into heat or other energy owing to friction and is dispersed. This form of damping corresponds to structural damping.

It is easy to realize from these examples that the viscous damping is proportional to the velocity of the node and the structural damping is proportional to the difference of the velocity between nodes. Let us assume that the damping matrix $\boldsymbol{C}$ consists of the viscous damping matrix $\boldsymbol{C}_v$ and the structural damping matrix $\boldsymbol{C}_s$.

$$\boldsymbol{C} = \boldsymbol{C}_v + \boldsymbol{C}_s. \tag{2.33}$$

We can anticipate that $\boldsymbol{C}_v$ will take the same form as the concentrated mass matrix and $\boldsymbol{C}_s$ the stiffness matrix. This assumption is based on the fact that the mass matrix is a coefficient constructed with regard to the acceleration of the node and the stiffness matrix is a coefficient constructed with regard to the difference between the nodal displacements.

In the case of a single element with two-nodes and one-degree of freedom, for instance, (2.33) it will be expressed as

$$\begin{bmatrix} c_{11} & c_{12} \\ c_{21} & c_{22} \end{bmatrix} \longrightarrow \begin{bmatrix} c_v & 0 \\ 0 & c_v \end{bmatrix} + \begin{bmatrix} c_s & -c_s \\ -c_s & c_s \end{bmatrix}.$$

Putting

$$\boldsymbol{C}_v = \alpha \boldsymbol{M}, \quad \boldsymbol{C}_s = \beta \boldsymbol{K}, \tag{2.34}$$

where $\alpha, \beta$: scalar constants, equation (2.33) can be expressed as

$$\boldsymbol{C} = \boldsymbol{C}_v + \boldsymbol{C}_s = \alpha \boldsymbol{M} + \beta \boldsymbol{K}. \tag{2.35}$$

Inserting this result into (2.32), we finally get the equation of motion

$$\boldsymbol{M}(\ddot{\boldsymbol{u}} + \alpha \dot{\boldsymbol{u}}) + \boldsymbol{K}(\beta \dot{\boldsymbol{u}} + \boldsymbol{u}) = \boldsymbol{P}. \tag{2.36}$$

This shows that the damping coefficient matrix can be expressed as a linear combination of the mass and the stiffness matrix.

### 2.3.2 Natural Frequency

Let us start our discussion with a system of one-degree of freedom. In the equation of natural vibration (1.7), we assume the displacement $u$ is a trigonometric function

$$u = u_0 e^{(i\omega t)}. \tag{2.37}$$

The velocity and the acceleration are obtained as follows

$$\dot{u} = i\omega u_0 e^{(i\omega t)} = i\omega u, \tag{2.38}$$
$$\ddot{u} = -\omega^2 u_0 e^{(i\omega t)} = -\omega^2 u \tag{2.39}$$

and (1.7) is expressed as an algebraic equation

$$(k - m\omega^2)u = 0. \tag{2.40}$$

For the existence of solution $u \neq 0$, the expressions within the parenthesis should be zero, or

$$\omega = \sqrt{\frac{k}{m}}.$$

The quotient $\omega/2\pi$ is called the natural frequency.

The same discussion can be developed in a system of multi-degrees of freedom. In the case of (1.10), the equation which relates to the algebraic equation (2.40) is

$$(\boldsymbol{K} - \lambda \boldsymbol{M})\boldsymbol{u} = 0. \tag{2.41}$$

Obtaining the scalar $\lambda$ and the vector $\boldsymbol{u}$ satisfying (2.41), the function

$$\boldsymbol{u} = \boldsymbol{u}_0 e^{(i\omega t)} \tag{2.42}$$

becomes the solution of the differential equation (1.10). Here $\omega = \sqrt{\lambda}$. The vector $\boldsymbol{u}_0$, which indicates the pattern of vibration, is called the mode of natural vibration. In the case that the right side of (2.32) is zero vector, i.e.,

$$\boldsymbol{M}\ddot{\boldsymbol{u}} + \boldsymbol{C}\dot{\boldsymbol{u}} + \boldsymbol{K}\boldsymbol{u} = \boldsymbol{0}, \tag{2.43}$$

the algebraic equation is expressed as

$$(\mu^2 \boldsymbol{M} + \mu \boldsymbol{C} + \boldsymbol{K})z = \boldsymbol{0}. \tag{2.44}$$

The eigenvalue and the eigenvector are converted to the complex number and the complex vector, respectively.

**(1) Standard eigenvalue problems** The eigenvalue problems taught at the introductory level of mathematics in colleges means that the scalar $\lambda$ that has a relationship

$$\boldsymbol{A}\boldsymbol{x} = \lambda \boldsymbol{x}, \tag{2.45}$$

where $\boldsymbol{A}$ is a square matrix. This eigenvalue has somewhat different meaning from the eigenvalue in the vibration problem. In this book, these eigenvalues shall be distinctively classified. The problem of (2.45) is called the standard eigenvalue problem and the problems of (2.41) and (2.43) are called the generalized eigenvalue problem.

The difference between two eigenvalues can be explained based on the geometrical meaning of vectors as follows. The general vector $\boldsymbol{x}$ converts to vector $\boldsymbol{y}$ with different magnitude and orientation by transformation $\boldsymbol{A}$. The standard eigenvector does not change the orientation of the vector by this transformation. Vector $\boldsymbol{u}$ in (2.41) does not change its orientation through two transformations $\boldsymbol{M}$ and $\boldsymbol{K}$.

The question before us is the relation between the standard eigenvalue problem and the generalized eigenvalue problem. Only the conclusion will be presented here. Working under the assumption that $\boldsymbol{M}$, $\boldsymbol{C}$, and $\boldsymbol{K}$ are all symmetric

(2.44) can result in (2.41) in a case of proportional damping,
(2.41) can result in (2.45), but its coefficient. $\boldsymbol{A}$ is not symmetric.

**(2) Orthogonality** For the existence of solutions in (2.41) and (2.44), the equations will be;

$$\det(\boldsymbol{K} - \lambda \boldsymbol{M}) = 0, \tag{2.46}$$

$$\det(\boldsymbol{K} + \mu \boldsymbol{C} + \mu^2 \boldsymbol{M}) = 0. \tag{2.47}$$

These matrixes are higher order equations of $\lambda$ and $\mu$, and are called the characteristic equation. The solution of the characteristic equation is an eigenvalue and solving simultaneous linear equations by the insertion of the eigenvalue into the original differential equation, we get the eigenvector corresponding to the eigenvalue.

Generally speaking, the eigenvalue and the eigenvector thus obtained take the following forms:

$\lambda$ is a real number and $\boldsymbol{u}$ is a real vector,
$\mu$ is a complex number and $\boldsymbol{z}$ is a complex vector.

Assuming the proportional damping, $\boldsymbol{z}$ becomes a real vector from the above reason.

In the standard eigenvalue problem, eigenvectors $\boldsymbol{x}_i$ and $\boldsymbol{x}_j$ corresponding to the different eigenvalues are in an orthogonal relation, i.e.

$$\boldsymbol{x}_i^T \boldsymbol{x}_j = 0 \tag{2.48}$$

In the generalized eigenvalue problem, the same relation exists when $\boldsymbol{M}$, $\boldsymbol{C}$ and $\boldsymbol{K}$ are all symmetrical matrixes.

In the case of (2.41), eigenvectors $\boldsymbol{u}_i$ and $\boldsymbol{u}_j$ corresponding to two eigenvalues $\lambda_i$ and $\lambda_j$ $(i \neq j)$ have the following relations

$$\boldsymbol{M} \text{ orthogonality:} \ \boldsymbol{u}_i^T \boldsymbol{M} \boldsymbol{u}_j = \boldsymbol{0}, \tag{2.49}$$

$$\boldsymbol{K} \text{ orthogonality:} \ \boldsymbol{u}_i^T \boldsymbol{K} \boldsymbol{u}_j = \boldsymbol{0}. \tag{2.50}$$

Assuming the proportional damping in (2.43), we will have $\boldsymbol{M}$ orthogonality and $\boldsymbol{K}$ orthogonality concerning eigenvectors $\boldsymbol{z}_i$ and $\boldsymbol{z}_j$, since (2.44) can result in (2.41) as was previously explained. This concept of orthogonality is one of the important properties of the eigenvector and constitutes a powerful clue when computing the eigenvalue. However, further discussion is out of the scope of this book. Particular discussion shall be left to other FEM books.

**(3) Modal Analysis** The method to analyze the forced vibration using the result of eigenvalue analysis in free vibration is called the modal analysis. Before entertaining into the main subject, the concept of modal matrix will be introduced.

First, we recast the eigenvalue problem of (2.41) as

$$\boldsymbol{K}\boldsymbol{x} = \lambda \boldsymbol{M}\boldsymbol{x}. \tag{2.51}$$

Next, we take $n$ eigenvectors which belong to the above equation. Assuming $\boldsymbol{M}$ orthogonality expressed by the following normalization

$$\boldsymbol{x}^T \boldsymbol{M} \boldsymbol{x} = 1. \tag{2.52}$$

The modal matrix is defined as

$$\boldsymbol{X} = \left[\, x_1 \,|\, x_2 \,|\cdots|\, x_n \,\right], \tag{2.53}$$

which involve the normalized $\boldsymbol{M}$ orthogonal vectors as its column vectors. The eigenvalue corresponding to $\boldsymbol{x}_i$ is noted by $\lambda_i$ $(= \omega_i^2)$.

Multiplying both sides of the differential equation of nondamping forced vibration

$$\boldsymbol{M\ddot{u}} + \boldsymbol{Ku} = \boldsymbol{P}(t) \tag{2.54}$$

by the transposition of the modal matrix $\boldsymbol{X}^T$ from left-hand side and carrying out the transformation of variables

$$\boldsymbol{u} = \boldsymbol{Xy},$$

we get an equation of $\boldsymbol{y}$ as follows;

$$\boldsymbol{X}^T\boldsymbol{MX\ddot{y}} + \boldsymbol{X}^T\boldsymbol{KXy} = \boldsymbol{G}(t). \tag{2.55}$$

Where

$$\boldsymbol{G}(t) = \begin{Bmatrix} g_1 \\ g_2 \\ \vdots \\ g_n \end{Bmatrix} \equiv \boldsymbol{X}^T\boldsymbol{P}(t). \tag{2.56}$$

Taking $\boldsymbol{M}$ orthogonality, $\boldsymbol{K}$ orthogonality, and (2.52) into consideration, (2.55) will be

$$\begin{bmatrix} 1 & 0 & \cdots & 0 \\ 0 & 1 & \cdots & 0 \\ \vdots & & & \vdots \\ 0 & \cdots & 0 & 1 \end{bmatrix} \ddot{\boldsymbol{y}} + \begin{bmatrix} \lambda_1 & 0 & \cdots & 0 \\ 0 & \lambda_2 & \cdots & 0 \\ \vdots & & & \vdots \\ 0 & \cdots & 0 & \lambda_n \end{bmatrix} \boldsymbol{y} = \begin{Bmatrix} g_1 \\ g_2 \\ \vdots \\ g_n \end{Bmatrix}. \tag{2.57}$$

In practical computations, it is sufficient to individually solve $n$ equations of one-degree-of-freedom system as follows;

$$\ddot{y}_i + \lambda_i y_i = g_i(t). \tag{2.58}$$

When the forced vibration involves the damping, a similar transformation is to be carried out on (2.32). Damping term will be added to (2.55) as follows;

$$\boldsymbol{X}^T\boldsymbol{MX\ddot{y}} + \boldsymbol{X}^T\boldsymbol{CX\dot{y}} + \boldsymbol{X}^T\boldsymbol{KXy} = \boldsymbol{G}(t) \tag{2.59}$$

Again working on the assumption of proportional damping, the damping coefficient $\boldsymbol{C}$ satisfies the orthogonality as it can be decomposed into $\boldsymbol{M}$ and $\boldsymbol{K}$ as is shown in (2.35), and the coefficient matrix of the second term in (2.59) will be a diagonal matrix as follows;

$$\boldsymbol{X}^T\boldsymbol{CX} \equiv \begin{bmatrix} \gamma_1 & 0 & \cdots & 0 \\ 0 & \gamma_2 & \cdots & 0 \\ \vdots & & & \vdots \\ 0 & 0 & \cdots & \gamma_n \end{bmatrix}.$$

In the practical computation, it is sufficient, as in the case of nondamping forced vibration, to solve individually n equations of one-degree-of-freedom system as follows;

$$\ddot{y}_i + \gamma \dot{y}_i + \lambda_i y_i = g_i. \tag{2.60}$$

### 2.3.3 Simulation

A static phenomenon is described as an equation of field, i.e., a partial differential equation of space. To treat this in FEM, the domain is divided into elements and polynomial approximation is done on each element. The analysis shall be carried out by solving these simple algebraic equations.

In a case of dynamic problems, on the other hand, time is added as an independent variable and the governing equation takes the form of a partial differential equation of time and space. The treatment of time is the key point in a dynamic problem.

In the modal analysis described in the previous paragraph, the displacement was assumed to be a trigonometrical function of time and treated semi-analytically.

Unfortunately, it is difficult to regard the response to, say, an impact force, as a trigonometrical function. It is only natural that we would seek a more direct and universal way of solving.

Now, let us recall the increment introduced in paragraph 2.2.1. Is there any possibility of relating this concept of increment to the time axis? The method of pursuing phenomenon by the time increment is called a simulation.

It is a numeric experiment by computer. The scope of application of the modal analysis was limited to only the equation of constant coefficients (2.32). In the case of simulation, however, it does not matter whether the coefficient is a function of time or the equation itself is nonlinear. It is a very universal method in such a meaning, however it has some faults like long analyzing time and poor analytical ability with limited response to a specific input. In short, the simulation exerts its full ability only in complex cases that exceed the ability of the modal analysis.

**(1) Euler's method** There are many practical methods used for simulation: the Runge-Kutta method, linear acceleration method, Newmark $\beta$-method, etc. Euler's method is the base of all these methods.

We will analyze (1.7), as an example, of Euler's method. Let $u_i$ and $\dot{u}_i$ denote the displacement and the velocity at time $t = t_i$, respectively. Then those at time $t = t + \Delta t$ shall be denoted by $u_j$ and $\dot{u}_j$, respectively. The displacement and the velocity can be simply expressed in the form

$$u_j = u_i + \dot{u}_i \Delta t,$$
$$\dot{u}_j = \dot{u}_i + \ddot{u}_i \Delta t.$$

From (1.7), we get

$$\ddot{u}_j = -m^{-1} k u_j.$$

Starting such procedures from the initial condition and proceeding stepwise for every $\Delta t$, we will get the solutions successively. The above example can be considered as a simulation of a specific node point discretized in FEM. The same process could be pursued in a system of multiple degrees of freedom, as in the case of (2.32).

$$\boldsymbol{u}_j = \boldsymbol{u}_i + \dot{\boldsymbol{u}}_i \Delta t, \tag{2.61}$$

$$\dot{\boldsymbol{u}}_j = \dot{\boldsymbol{u}}_i + \ddot{\boldsymbol{u}}_i \Delta t, \tag{2.62}$$

$$\ddot{\boldsymbol{u}}_j = \boldsymbol{M}^{-1}(\boldsymbol{P}_j - \boldsymbol{K}\boldsymbol{u}_j - \boldsymbol{C}\dot{\boldsymbol{u}}_j). \tag{2.63}$$

**(2) Linear acceleration method** The linear acceleration method is an improved version of the Euler's methods. This is literally a method that approximates the change of acceleration by linear approximation. For example, speed is expressed in one-degree of freedom,

$$\dot{u}_j = \dot{u}_i + \frac{1}{2}(\ddot{u}_i + \ddot{u}_j)\Delta t. \tag{2.64}$$

The displacement is, taking up to the third-order term of a Taylor expansion, approximated by

$$u_j = u_i + \dot{u}_i \Delta t + \ddot{u}_i \frac{(\Delta t)^2}{2!} + \frac{(\ddot{u}_j - \ddot{u}_i)}{\Delta t} \frac{(\Delta t)^3}{3!}. \tag{2.65}$$

In the actual procedure, we specify simultaneous equations (2.64), (2.65), and the vibration equation in concern. We then seek the solution by following step-by-step procedure.

For further explanation, please refer to [3].

## 2.4 SPATIAL DISCRETIZATION

### 2.4.1 Derivating Procedure of Element Stiffness

The outline of spatial discretization was described in section 1.3 and some concrete examples such as the construction of the global stiffness from the element stiffness and the treatment of boundary conditions were explained in detail. In paragraph 1.5.2 we concerned ourselves with elastic components and in paragraph 1.5.3 we discussed plane elements. The explanation of Step 1 was omitted in chapter 1 because the element stiffness was assumed to be known quantity. Step 2, ..., Step 4 of paragraph 1.3.2 were explained previously.

In a general elastic body, elastic stiffness is unknown and the (Step 1) procedure is indispensable. Now, a more detailed procedure of (Step 1) will be presented here on displacement method. Since displacement was designated as the final variable in the displacement method, nodal displacement shall be expressed after discretization. i.e., The principle of virtual work by the nodal displacement should be pursued eliminating stress or strain from the basic equation. To this end, we follow next steps.

1. Denote the displacement within the element by the nodal displacement: Assume the displacement function and substitute the nodal coordinate into the displacement function.

2. Denote the strain by the nodal displacement: Substitute the result of 1. into strain/displacement equation.

3. Denote the stress by the nodal displacement: Substitute the result of 2. into stress/strain equation.

4. Express the principle of virtual work by the nodal displacement: We obtain the equation equivalent to the equilibrium equation.

The key point in the above steps is the selection of the displacement function in step 1. The approximation of the variable function by the lower order equation is the fundamental procedure of FEM. The element used in the 1st order approximation is referred to as a linear element and a quadratic element in the case of a 2nd order approximation. The accuracy of the solution can be improved, through the use of higher order equations. When the discretization is carried out in exacting detail, the accuracy will be greatly improved, even if a lower order equation is used.

Steps 2, ..., 4 are mechanical and substitutive operations that require little effort. These operations are easily understood on some comprehensive examples. The derivation of the stiffness will be demonstrated hereafter on some popular cases following steps 1, ..., 4. The type of the element is assumed to be linear.

### 2.4.2 Stiffness of Truss Elements

Although the reader would have learned the derivation of (1.3) in an elementary course of study, we derive it from the basic equation of section 2.1, following the steps of FEM. As to the meaning of notations, refer to Fig. 2.1.

1. Denote the displacement within the element by the nodal displacement:

First, we approximate the displacement function by a linear equation of $x$

$$u(x) = \{ 1 \quad x \} \begin{Bmatrix} \alpha_1 \\ \alpha_2 \end{Bmatrix}. \tag{2.66}$$

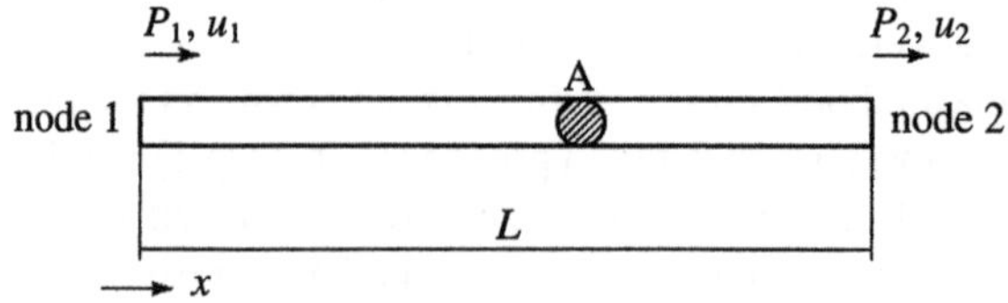


Figure 2.1: A truss element

Substituting the nodal coordinate and the nodal displacement into the above equation, we get

$$\begin{Bmatrix} u_1 \\ u_2 \end{Bmatrix} = \begin{bmatrix} 1 & x_1 \\ 1 & x_2 \end{bmatrix} \begin{Bmatrix} \alpha_1 \\ \alpha_2 \end{Bmatrix}.$$

Solving this equation, we can determine the undetermined coefficients $\alpha_1$ and $\alpha_2$

$$\begin{Bmatrix} \alpha_1 \\ \alpha_2 \end{Bmatrix} = \frac{1}{L} \begin{bmatrix} x_2 & -x_1 \\ -1 & 1 \end{bmatrix} \begin{Bmatrix} u_1 \\ u_2 \end{Bmatrix}, \tag{2.67}$$

where $L$ is the length of a structural member $L = x_2 - x_1$.

Substituting (2.67) into $\alpha$ of (2.66), we can express the displacement function by the nodal displacement.

2. Denote the strain by the nodal displacement:

Differentiating (2.66) with respect to $x$, we get the normal strain

$$\varepsilon_x = \frac{\partial u}{\partial x} = \frac{1}{L}\{\, 0 \quad 1 \,\} \begin{bmatrix} x_2 & -x_1 \\ -1 & 1 \end{bmatrix} \begin{Bmatrix} u_1 \\ u_2 \end{Bmatrix} = \frac{1}{L}\{\, -1 \quad 1 \,\} \begin{Bmatrix} u_1 \\ u_2 \end{Bmatrix}. \tag{2.68}$$

The virtual strain corresponding to the virtual nodal displacement is,

$$\delta\varepsilon = \frac{1}{L}\{\, -1 \quad 1 \,\} \begin{Bmatrix} \delta u_1 \\ \delta u_2 \end{Bmatrix}. \tag{2.69}$$

3. Denote the stress by the nodal displacement:

Substituting

$$\varepsilon_y = \varepsilon_x = -\nu\varepsilon_x, \quad \gamma_{xy} = \gamma_{yz} = \gamma_{zx} = 0$$

into (2.9), we get from (2.68)

$$\sigma_x = E\varepsilon_x = \frac{E}{L}\{\, -1 \quad 1 \,\} \begin{Bmatrix} u_1 \\ u_2 \end{Bmatrix}. \tag{2.70}$$

4. Express the principle of virtual work by the nodal displacement:

$$\text{External virtual work} = \{\, P_1 \quad P_2 \,\} \begin{Bmatrix} \delta u_1 \\ \delta u_2 \end{Bmatrix},$$

$$\text{Internal virtual work} = \int_{x_1}^{x_2} (\sigma_x^T \cdot \delta\varepsilon_x) A\, dx = \frac{EA}{L}\{\, u_1 \quad u_2 \,\} \begin{bmatrix} 1 & -1 \\ -1 & 1 \end{bmatrix} \begin{Bmatrix} \delta u_1 \\ \delta u_2 \end{Bmatrix}.$$

Since these two equations are equivalent at any virtual displacement, we eventually get

$$\begin{Bmatrix} P_1 \\ P_2 \end{Bmatrix} = \frac{EA}{L} \begin{bmatrix} 1 & -1 \\ -1 & 1 \end{bmatrix} \begin{Bmatrix} u_1 \\ u_2 \end{Bmatrix}. \tag{2.71}$$

Compare the above procedure with the derivation of (1.12) using the principle of virtual work described in paragraph 1.3.4.

### 2.4.3 Element Stiffness of Plane Stress

Let us find the component of the element stiffness in (1.57), assuming the element shown in Fig. 1.21 is a plane element that deforms within a single plane. The element has a triangular form with node numbers shown in Fig. 2.2; the $x$ and the $y$ component of the displacement vector are denoted by $u$ and $v$, respectively.

The difference of this element from the truss is only increased degrees of freedom of the node. Follow the process paying attention to this point.

1. Denote the displacement within the element by the nodal displacement:

The displacement function shall be assumed to be a linear equation

$$\begin{Bmatrix} u \\ v \end{Bmatrix} = \begin{bmatrix} 1 & x & y & 0 & 0 & 0 \\ 0 & 0 & 0 & 1 & x & y \end{bmatrix} \begin{Bmatrix} \alpha_1 \\ \alpha_2 \\ \alpha_3 \\ \alpha_4 \\ \alpha_5 \\ \alpha_6 \end{Bmatrix}. \tag{2.72}$$

Substituting the nodal coordinate and the nodal displacement into (2.72), we get

$$\begin{Bmatrix} u_1 \\ v_1 \\ u_2 \\ v_2 \\ u_3 \\ v_3 \end{Bmatrix} = \begin{bmatrix} 1 & x_1 & y_1 & 0 & 0 & 0 \\ 0 & 0 & 0 & 1 & x_1 & y_1 \\ 1 & x_2 & y_2 & 0 & 0 & 0 \\ 0 & 0 & 0 & 1 & x_2 & y_2 \\ 1 & x_3 & y_3 & 0 & 0 & 0 \\ 0 & 0 & 0 & 1 & x_3 & y_3 \end{bmatrix} \begin{Bmatrix} \alpha_1 \\ \alpha_2 \\ \alpha_3 \\ \alpha_4 \\ \alpha_5 \\ \alpha_6 \end{Bmatrix}. \tag{2.73}$$

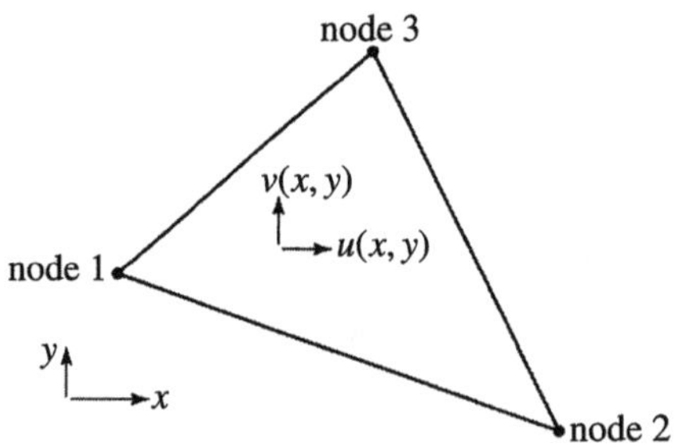


Figure 2.2: A triangular element

Solving this equation, we get the undetermined coefficients $\alpha_1, \ldots, \alpha_6$.

$$\begin{Bmatrix} \alpha_1 \\ \alpha_2 \\ \alpha_3 \\ \alpha_4 \\ \alpha_5 \\ \alpha_6 \end{Bmatrix} = \frac{1}{\Delta} \begin{bmatrix} \Delta_1 & 0 & \Delta_2 & 0 & \Delta_3 & 0 \\ 0 & \Delta_1 & 0 & \Delta_2 & 0 & \Delta_3 \\ y_2 - y_3 & 0 & y_3 - y_1 & 0 & y_1 - y_2 & 0 \\ 0 & y_2 - y_3 & 0 & y_3 - y_1 & 0 & y_1 - y_2 \\ x_3 - x_2 & 0 & x_1 - x_3 & 0 & x_2 - x_1 & 0 \\ 0 & x_3 - x_2 & 0 & x_1 - x_3 & 0 & x_2 - x_1 \end{bmatrix} \begin{Bmatrix} u_1 \\ v_1 \\ u_2 \\ v_2 \\ u_3 \\ v_3 \end{Bmatrix}, \quad (2.74)$$

where

$$\Delta = \begin{vmatrix} 1 & x_1 & y_1 \\ 1 & x_2 & y_2 \\ 1 & x_3 & y_3 \end{vmatrix} : \text{Area of triangle } S = |\Delta|/2,$$

$$\Delta_1 = \begin{vmatrix} x_2 & x_3 \\ y_2 & y_3 \end{vmatrix}, \quad \Delta_2 = \begin{vmatrix} x_3 & x_1 \\ y_3 & y_1 \end{vmatrix}, \quad \Delta_3 = \begin{vmatrix} x_1 & x_2 \\ y_1 & y_2 \end{vmatrix}.$$

Substituting (2.74) into $\alpha$ of (2.72), the displacement function can be expressed by the nodal displacements.

2. Denote the strain by the nodal displacement:

Differentiating (2.72) with respect to $x$, non-zero component of the strain shall be

$$\begin{Bmatrix} \varepsilon_x \\ \varepsilon_y \\ \gamma_{xy} \end{Bmatrix} = \frac{1}{\Delta} \begin{bmatrix} y_2 - y_3 & 0 & y_3 - y_1 & 0 & y_1 - y_2 & 0 \\ 0 & x_3 - x_2 & 0 & x_1 - x_3 & 0 & x_2 - x_1 \\ x_3 - x_2 & y_2 - y_3 & x_1 - x_3 & y_3 - y_1 & x_2 - x_1 & y_1 - y_2 \end{bmatrix} \begin{Bmatrix} u_1 \\ v_1 \\ u_2 \\ v_2 \\ u_3 \\ v_3 \end{Bmatrix}. \quad (2.75)$$

We denote the equation by $\boldsymbol{\varepsilon} = \boldsymbol{Bu}$ for the sake of simplicity. Then the virtual strain corresponding to the virtual nodal displacement will be given by $\delta\boldsymbol{\varepsilon} = \boldsymbol{B}\delta\boldsymbol{u}$.

3. Denote the stress by the nodal displacement: Extracting the strain component corresponding to (2.75) from (2.9), we get

$$\begin{Bmatrix} \sigma_x \\ \sigma_y \\ \tau_{xy} \end{Bmatrix} = \frac{E}{1 - \nu^2} \begin{bmatrix} 1 & \nu & 0 \\ \nu & 1 & 0 \\ 0 & 0 & (1-\nu)/2 \end{bmatrix} \begin{Bmatrix} \varepsilon_x \\ \varepsilon_y \\ \gamma_{xy} \end{Bmatrix}. \quad (2.76)$$

Denoting the equation by $\boldsymbol{\sigma} = \boldsymbol{D\varepsilon}$, we get

$$\boldsymbol{\sigma} = \boldsymbol{DBu}. \quad (2.77)$$

4. Denote the principle of virtual work by the nodal displacement: Let $P_i$ and $Q_i$ $(i = 1, 2, 3)$ denote the $x$ and the $y$ component of the nodal force, respectively, we

have the following expressions

$$\text{Exterior virtual work} = \{\ P_1 \ \ Q_1 \ \ P_2 \ \ Q_2 \ \ P_3 \ \ Q_3\ \} \begin{Bmatrix} \delta u_1 \\ \delta v_1 \\ \delta u_2 \\ \delta v_2 \\ \delta u_3 \\ \delta v_3 \end{Bmatrix} \equiv \boldsymbol{P}^T \delta \boldsymbol{u},$$

$$\text{Interior virtual work} = \iint_S (\boldsymbol{\sigma}^T \delta \boldsymbol{\varepsilon}) t \, dxdy = S\, t \boldsymbol{u}^T \boldsymbol{B}^T \boldsymbol{D} \boldsymbol{B} \delta \boldsymbol{u},$$

where $t$ denotes the thickness of the plate. The symmetry of $\boldsymbol{D}$ is considered.

Since these two equations are equivalent at any virtual displacement, we eventually get

$$\boldsymbol{P} = (S\, t \boldsymbol{B}^T \boldsymbol{D} \boldsymbol{B}) \boldsymbol{u}. \tag{2.78}$$

The inside of the parenthesis ( ) constitutes an element stiffness matrix.

### 2.4.4 Element Stiffness of a Three-dimensional Elastic Body

We will expand the current discussion into a three-dimensional elastic body shown in Fig. 2.3.

We express the displacement function within the element as

$$\boldsymbol{u} = \boldsymbol{u}(x, y, z) = \{\ u \ \ v \ \ w\ \}^T,$$

and denote the nodal displacement and the nodal force by $\boldsymbol{U}$ and $\boldsymbol{P}$, respectively. The number of components of $\boldsymbol{U}$ and $\boldsymbol{P}$ is three times as many as the number of nodes.

1. Denote the displacement within the element by the nodal displacement:

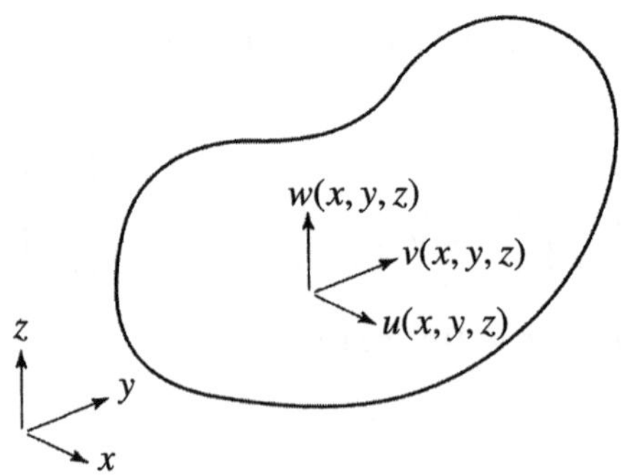


Figure 2.3: A three-dimensional elastic body

Approximating the displacement function by the linear function of $x$, $y$, and $z$. The coefficient $\boldsymbol{M}$ of

$$\boldsymbol{u} = \boldsymbol{M\alpha}, \tag{2.79}$$

is given by a linear function of $x$, $y$, and $z$. Substituting the nodal coordinate and the nodal displacement into the above equation and solving for the unknown coefficient $\boldsymbol{\alpha}$, we get the expression in the same way as (2.74)

$$\boldsymbol{\alpha} = \boldsymbol{TU}, \tag{2.80}$$

where $\boldsymbol{T}$ is a constant coefficient. Substituting (2.80) into (2.79)

$$\boldsymbol{u} = \boldsymbol{MTU}, \tag{2.81}$$

the displacement function has been expressed by the nodal displacement.

2. Denote the strain by the nodal displacement:

Substituting (2.79) into (2.8), the result will be

$$\boldsymbol{\varepsilon} = \boldsymbol{G\alpha}$$

by simplified expression. Substituting (2.80) into the above expression, we get

$$\boldsymbol{\varepsilon} = \boldsymbol{GTU} \equiv \boldsymbol{BU}. \tag{2.82}$$

The virtual strain corresponding to the virtual nodal displacement will be expressed as $\delta\varepsilon = \boldsymbol{B}\delta\boldsymbol{U}$.

3. Denote the stress by the nodal displacement:

Substituting (2.82) into (2.9), we get

$$\boldsymbol{\sigma} = \boldsymbol{D\varepsilon} = \boldsymbol{DBU}.$$

4. Express the principle of virtual work by the nodal displacement:

Equating the following two terms

$$\text{Exterior virtual work} = \boldsymbol{P}^T \delta\boldsymbol{U},$$

$$\text{Interior virtual work} = \iiint_V (\boldsymbol{\sigma}^T \delta\boldsymbol{\varepsilon})\, dxdydz = \boldsymbol{U}^T \iiint \boldsymbol{B}^T \boldsymbol{DB}\, dxdydz\delta\boldsymbol{U},$$

and taking the arbitrariness of the virtual displacement into account, we finally get

$$\boldsymbol{P} = \iiint (\boldsymbol{B}^T \boldsymbol{DB})\, dxdydz\boldsymbol{U} \equiv \boldsymbol{kU}. \tag{2.83}$$

Eventually the element stiffness matrix of a three-dimensional elastic body has been obtained as $\boldsymbol{k}$ in (2.83).

## Bibliography of Chapter 2

[1] Yamada Y, Plastic and Viscoelastic Materials (in Japanese), Baifukan (1980)
[2] Togawa H, Vibration Analysis by FEM (in Japanese), Science Press (1975), p23
[3] Shimoseki and Fujinuma, Practical Programming for FEM (in Japanese), Nikkan Kogyo Press (1989), p63

# 3 ROLE OF FEM IN SPRING ANALYSIS

Generally speaking, the analytical approach for spring calculations has been fully developed. Further effort to bring about new calculation equations would not result in additional benefit. Therefore, it seems that only numerical analysis is left to us to improve the accuracy. There are many levels in the conventional analysis technique, and the necessity for selecting the appropriate formula is attributed to the limited area of application of each formula. We will examine the case of a helical compression spring.

The design formula of the helical spring (3.1) has been given in many handbooks and should be familiar to our readers. These formulas are constructed on an assumption that a model of one-degree of freedom replaces the actual spring as an elastic component described in chapter 1. In this case, no degree of freedom is allowed in the transversal direction and we cannot get the transverse stiffness ($\kappa$ in subsection 1.4.5). Now, what is the usual way for finding the transversal stiffness of actual springs? The makeshift formula for the transversal stiffness[1] could be used. This formula is nothing more than an elastica, which is a model with an increased degree of freedom in an elastic component. It naturally has a limit in its application. For example, it is natural that the transversal stiffness has its own direction that originates from relative spring end position and the direction of transverse force. But, it is fundamentally impossible to consider the direction of the transversal stiffness as long as the spring is treated as an elastica.

When we want to analyze such details of the spring, the only method is to formulate the spring as a curved beam with three-dimensional six-degrees of freedom. There are several such outcomes[2, 3]. While it is possible to analyze the direction of the transversal stiffness or the eccentricity of force using these formulas, the problem is the laborious calculation. While solving can be carried out by hand for simple formulas, computer is essential to make calculation on much complicated formulas.

Although the formula appears to be the best way for analysis, it has its own limits due to its analytical character. That is, the shape of the spring must be expressed mathematically; meaning the shape of the spring is limited to something fairly primitive. The suspension spring for passenger cars, for instance, has a very complicated shape. It might be conically wound on the ends in some cases. It might even be tapered along the length of the wire. The conventional calculation cannot cope with such a complicated shape. Then, the conception of FEM of dividing the spring into elements was inevitably formulated[4]. In the early days, however, we did nothing but make the computer program ourselves.

This situation was suddenly changed when the commercial FEM program became popular. Even if we cannot make a computer program, it is possible to carry out computations with proper sequence of data input, when the FEM software is used as a black box. The popularization of such techniques has liberated the engineer from laborious calculations.

In reviewing the history of analysis for the helical spring, the reader might have noticed that the analytical technique has arrived at the final stage of its improvement. Looking at springs of other types, we unfortunately find many springs that are still at the level of hand calculation and have not benefited from recent improvements in analysis. Flat springs and formed wire springs have been left almost untouched. This cannot be attributed, however, to the negligence of the engineer. These springs are so diversified in their shapes, that even a simplified equation could not be formulated thus far. There was nothing else to do but apply Castigliano's theorem to each case and to establish an appropriate design formula for the particular shape. For springs of these types, the appearance of FEM beyond doubt, an absolute lifesaver. FEM does not need the design formula, and shows its proper function in an a-la-carte menu computation.

What the reader may want to know might not be the content of FEM, but its usefulness in daily works. It is the main objective of this chapter to show the merit of FEM for spring technique. In a sense, this is a kind of tour to visit the greenroom of spring theater.

## 3.1 COMPARISON OF FEM WITH CONVENTIONAL DESIGN METHODS

### 3.1.1 Assumption in Model Construction

**(1) Simplified formula for helical springs** The most popular calculating formula of springs are those for cylindrical compression springs with circular cross-sectional wire. With following notation,

$k$ = spring constant, $d$ = wire diameter, $D$ = mean coil diameter,

$n$ = effective number of coils, $G$ = shear modulus,

$\tau$ = shear stress (modified), $\kappa$ = curvature correction factor,

the formulas are given by

$$\text{spring constant: } k = \frac{Gd^4}{8nD^3}, \tag{3.1}$$

$$\text{surface stress: } \tau = \kappa\frac{8PD}{\pi d^3}. \tag{3.2}$$

Equation (3.1) specifies spring characteristic, and (3.2) indicates spring strength.

The assumptions used in the derivation of (3.1) are:

- Disregard the effect of a spatially located curved beam .......... The model is built as the twisting of a straight bar.

- External force is applied along the central axis of the coil .......... Eccentric oblique loads or moments from the ends are disregarded.

- Deformation is assumed to be minute .......... Deflection is regarded as negligibly small compared to the dimension of the spring.

Equation (3.2), on the other hand, takes the curvature into consideration while leaving the other assumptions as they are. It seems that the formula is standing on the assumption having no consistency. This is the familiar modified stress by Wahl.

These formulas could be used in the majority of cases. But it is irrelevant to make a complaint about the inaccuracy of the formula in such cases as:

- A large pitch angle; Disregard of the bending moment cannot be allowed.

- A small number of n; The effect of the eccentric oblique load cannot be disregarded.

- Higher stress on the outside diameter than on the inside diameter;

- Uneven distribution of the stress over the surface of the coils;

It is only natural that each formula would have its own limits in its application, depending on the imposed assumptions.

As is described above, the more accurate the model must be, the more complicated mathematical equations should be. It is a great comfort for us to shift to FEM, rather than struggle with complicated equations.

**(2) Tip contact method for a multiple leaf spring** As for multiple leaf springs, we have a very interesting calculation technique called the tip contact method. The tip contact method is a forerunner of FEM and is worth mentioning for the viewpoint of this book. The assumptions on which the method was constructed are:

- Each leaf contacts at both ends and at the center of leaf;

- Leaves are assumed to be straight beams;

- Deformation is assumed as being minute.

The method treats nothing but a model in which each leaf is divided into two at the contact position and the resulting two beam elements are used as the members to be analyzed. Besides, the global stiffness matrix is tridiagonal from the first assumption, and hand calculation could be performed within tolerable effort, making this method attractive.

In this model, the relative slippage between leaves won't be handled because the camber of the leaf is neglected, and minute deformation is assumed in this solution.

Where no slippage exists, no friction is anticipated.

This means that the consideration of friction in the tip contact method is fundamentally inadequate. Even so, it is also true that such almost reckless treatments have been reported in the past. The arrangement of the problem into the matrix expression results in a straightforward appearance of mathematical equations, and helps us to have a clear view of the theoretical ground, adding more merit to the method. It would be worth reevaluating past reports on multiple leaf springs at this time; The discussion will be added to the introduction of concrete examples in chapters 5 and 6.

### 3.1.2 From Linear to Nonlinear

From the beginning, it should be noted that FEM is not necessary unless we feel inconvenient by conventional methods. The following discussion will be developed on an assumption that we have a kind of dissatisfaction with conventional calculation formulas. Then it is necessary to see if the problem can be solved by FEM and, when it is applicable, how hard the work will be. When the analysis is classified by linearity, the analysis of the linear type is easier than the nonlinear analysis. We can assume that all spring problems, including dynamic ones, can be solved by FEM, as long as the deformation is assumed to be small. As most conventional design formulas are based on linear analysis, any problem that can be solved with the conventional method will also be solvable with FEM. There are still many unsolved problems by conventional methods, even if they belong to the linear analysis category. Examples of such linear analysis will be introduced in chapter 5.

The nonlinear problem is a tough job. Even so, the analysis is possible in principle. Some cases that need to take the nonlinearity into account in the actual spring design will be discussed here. As stated previously, there are two sorts of nonlinearity; geometrical nonlinearity and material nonlinearity. Having this in mind, we could say the following:

| | |
|---|---|
| Manufacturing of springs: | The geometrical and the material nonlinearity will come into play. |
| Spring products: | Only the geometrical nonlinearity will come into play. |

As long as the characteristics or the strength of product are considered, the product is assumed to be an elastic body and there is no need to consider the nonlinearity of material except in a special circumstance like creep. However, the deformation of springs are, as is repeatedly stressed, larger than other mechanical components and the geometrical nonlinearity of spring often becomes an issue to be solved. If we want to specify the beginning point of contact or to take the friction of the multiple leaf spring into account, the problem should be treated as a geometrical nonlinearity case. If we are content with the level of the approximation not being much higher than the tip contact method, then the linear analysis is sufficient. However, the plastic deformation cannot be neglected when the manufacturing process is to be simulated. In this case, spring often shows geometrical nonlinearity, where the strain of spring is so large during the forming process that the spring deforms plastically. The prediction of permanent set in the pre-setting process corresponds to this case.

The reader has learned by now, that there are various levels of FEM to be introduced into the spring design, according to our requirements. Even though an advanced research is necessary for this reason, it is quite difficult in practice. It is hard work to translate an actual task into a mathematical model. Unless the main point of the problem is made clear, we cannot tell whether the task requires a linear or a nonlinear analysis. As a matter of fact, this decision is a key point with which our true technical ability is evaluated. A problem becomes as clear as it has already been solved when all points of concern are in order.

## 3.2 THE UTILIZATION OF FEM SOFTWARE

### 3.2.1 Use of Commercial Software

They say "Practice makes perfect." This applies to FEM as much as anything else in life. For the beginner, it is recommended to use existing software instead of developing a homemade program starting from scratch. Like other application software, our expertise will improve to add custom-made functions once we get accustomed to it.

For those not inclined to spend time familiarizing themselves with FEM, there is the way of a subcontractor. There are firms around the world whose occupation is to undertake the analysis of FEM. The author does not personally recommend this option. Expertise and experience would not be acquired.

Limiting the task to a linear problem, any commercial software could be run on a PC. When it comes to nonlinear problems, a PC is not adequate enough. The use of an EWS (engineering work station) is inevitable for a nonlinear problem. The capability of commercial software to handle nonlinear problems seems to be adequate in general.

Summing up the discussion, there is no big difference among commercially available FEM software. Tips on selecting commercial software is summarized as follows. Difference are minor.

**(1) Multifunction is not always favorable;** In most cases, only a small part of the entire software capability is needed for practical work. It may be a waste of money to purchase software fully equipped with the capability of nonlinear analysis only for a linear task. It may be a case that a few elements are needed from a lineup of several hundred. When the analysis requirements are determined, it may be wise to purchase module software geared towards a specific purpose[5].

**(2) Preprocessing (preliminary preparation) level;** The most troublesome aspect of preprocessing is the discretization into mesh, i.e., the input work of nodal coordinates and the definition of elements. Many functions are now being developed to reduce the burden of such jobs. The emphasis of software design seems to have shifted from FEM itself to the preprocessing software in market competition. The ideal preprocessing would be to import the drawing information from CAD (computer-aided design) data and then carry out discretization into mesh. Automatic discretization may be possible for flat springs or formed wire springs, which have their shapes uniquely determined by their drawing information. In the case of multiple leaf springs or helical springs, their drawings only show the representative dimensions, and they can not be utilized as input-data.

**(3) Availability of a user interface;** In a process to define, for instance, a helical spring by beam elements, it is agonizingly difficult to define the nodes on a spiral in a three-dimensional space. For a design engineer, it is preferable to employ a simplified formula. The technique of automatic discretization is not yet advanced to this level. The only way to cover this is that users construct a personal program by themselves to save the labor of input. Some analysts may even wish to remodel the post-processing software in ones own style. If a user wants to get the principal stress distributed along the inside diameter of the coil, it is inevitable for them to add an appropriate program. As such remodeling is anticipated, it is better to consider the availability of user interface in advance.

### 3.2.2 Selection of Commercial Software

More than 30 products of full-scale software for FEM analysis are available in Japanese domestic market. In addition, there are many kinds of preprocessing and post-processing software. Thinking about all combinations is beyond consideration. However, it is necessary to have some knowledge at hand when we examine the introduction of commercial software. Table 3.1 provides an evaluation of six representative software packages. Please be aware that the evaluations reflect opinions and biases of author.

For linear analysis, commercial software is good enough in the majority of cases. The difference in their ability emerges when treating nonlinear analysis. As has been repeatedly stressed, the nonlinear problem is very important in the analysis of springs. Therefore, the adaptability to three kinds of nonlinearity (boundary nonlinearity was classified separately) was selected as the benchmark for evaluation.

Table 3.1: FEM software performance table of nonlinear capacity

| | ADINA | I-DEAS | NASTRAN | ABAQUS | MARC | ANSYS |
|---|---|---|---|---|---|---|
| Geometrical nonlinearity | ○ | ○ | ○ | ◎ | ○ | ○ |
| Material nonlinearity | ○ | △ | △ | ○ | ◎ | △ |
| Boundary nonlinearity | ◎ | △ | △ | ◎ | ○ | △ |

◎ Excellent ○ Good △ Restricted

### 3.2.3 Development of Dedicated Programs

A comparatively easy problem of linear analysis can be successfully achieved by a personal or even with a pocket computer. It might be beneficial for readers to construct some programs for the purpose of learning and experience. Even though ready-made software is going to be introduced sooner or later, it is recommended to attempt the challenge of a handmade program at least once. The challenge will give the reader confidence of being able to look into the black box. Books on FEM that carry program lists are readily available in the market and will serve the reader as a reference.

When we are not content with ready-made software, that is, where we want to analyze a problem using odd elements which are not provided in the package, there is nothing to be done but construct the necessary element stiffness matrix by ourselves. Even in this case, ready-made software can be utilized in the process for the assemblage of global stiffness matrix and the solution of the simultaneous linear equations. As was described in paragraph 1.3.2, FEM has been standardized and it's possible to replace individual processing unit modules. Even in the case of developing a special purpose program, it may be practical to utilize the software at hand as much as possible, and then modify the program for other necessary requirements.

Some problems may not allow us to easily determine the boundary conditions. A suspension spring for passenger cars, for example, is being assembled in a link mechanism as is shown in Fig. 3.1.

The lower seat in the figure has a spiral slope and the end turn of the spring also has a spiral shape. In this case, the two three-dimensional spirals are contacting obliquely and the boundary condition cannot be determined unless contact point is specified. This could be treated basically by preprocessor, but sometimes it becomes impossible to analyze contact point under insecure condition. A supplemental strut

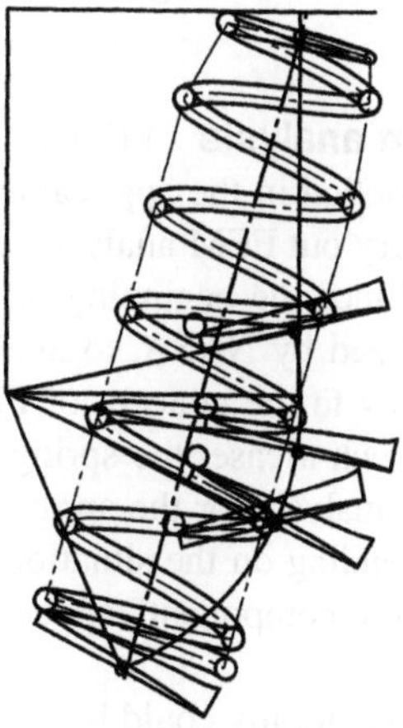

Figure 3.1: A spring incorporated in a link mechanism.

may be used to mount the actual spring in some cases, then innovative ideas are required to express this mechanism as a boundary condition. For such a problem, there is nothing to do but, construct a personal program[6].

## 3.3 EFFECTIVENESS IN DESIGN PRACTICE

### 3.3.1 Single Spring and Peripheral Parts

**(1) Verification of the design specification for springs** The first point to pay attention to, when using FEM for the design practice of springs, is that FEM is not a design tool but an analytical tool. The term "design" means to determine the shape and/or material so as to fulfill the functional and spatial requirements. An analysis is a practice entirely opposite to this, that is, the analysis is to verify spring characteristics or stresses under the given conditions of the material and shape. Therefore, FEM is useful in the final stages of the design. In determining specifications, we must treat so-called inverse problems. A computer program that can treat the inverse problem is on the market under the name of "Optimum Design Solver." It could shave off the excess amount of thickness where stress is well below allowable level, but an advanced task like the automatic decision of spring specifications is far from being performed. In the case of helical springs or multiple leaf springs, for which some simplified formulas exist, there is a compromise plan, namely to use the formulas for the initial rough estimation and FEM for the final confirmation. In the case of springs without a design formula, referring to, flat springs, formed wire springs, and so on, there is nothing to do but use common sense and experience.

Conventional design practice cannot be ignored even after the introduction of full scale FEM software.

**(2) Expansion to system analysis** FEM is just not a dedicated tool solely for springs. There is no restriction in the application of FEM. With this in mind, it might be a good idea to carry out FEM analysis on the whole system, including peripheral parts of springs. Since the beginning of development, the objective of FEM has been, as is symbolized by NASA, to analyze an object integrally from part to the whole system. Key to the concept of FEM is the freedom to enlarge the territory of analysis. In such a case, the spring is just a constituting part and there is no need of a precise analysis for the spring. One-degree of freedom may be enough for the spring depending on the situation. Spring element registered in commercial software or "elastic component" appeared in chapter 1 could be used for this purpose.

In many cases, the problem naturally could become more complicated, however, we can turn the problem into something simpler if we treat it skillfully. A good example of such a case will be presented here. It is an analysis on the behavior of the

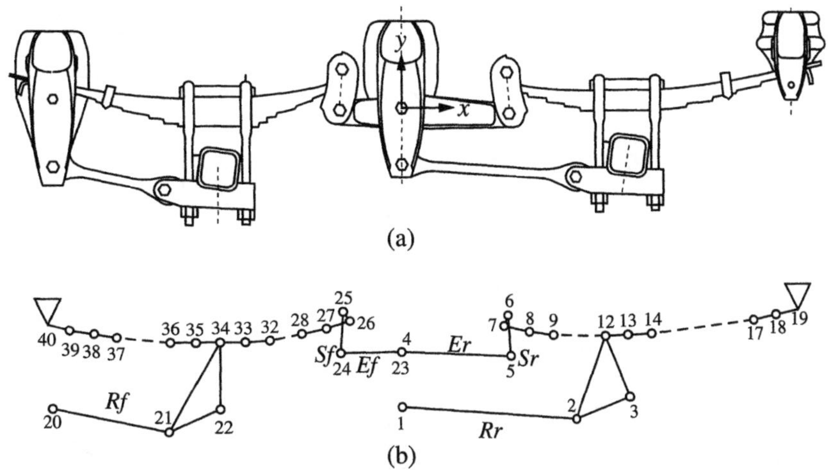


Figure 3.2: A model of a truck suspension mechanism.

whole suspension system of a truck, including multiple leaf springs, by the RBSM method[7]. RBSM is an abbreviation for RIGID-BODIES-SPRING-MODEL and is a zero-order element from the standpoint of FEM[8]. Fig. 3.2(a) shows a rear suspension of a two-row rear axle, and (b) is its model. Fig. 3.3(a) shows the deformation of the suspension when node 3 has run over a bump $\delta$ and (b) shows the normal force $Q$ at the position of the axles.

Even though the problem here is in large deformation situation, RBSM method can handle the solving process as linear analysis using bending moment parameter and turning angle parameter separately. The notations $Y_i$ and $y_i$ in Fig. 3.3(a) are the vertical coordinates of node $i$ before and after the deformation. Since the external forces ($E_r$, $E_f$, $S_r$, $S_f$, $R_r$, and $R_f$ ) acting on each structural member with the notation in Fig. 3.2 can be obtained as is shown in Fig. 3.3(c), we can use these forces, if necessary, as the mechanical boundary condition for each structural member and then the detailed stress analysis could be performed.

### 3.3.2 Simulation of the Manufacturing Process

The simulation of spring manufacturing process is not easy to handle since we must take spring back into account. Plastic deformation is always accompanied by spring back. In the case of springs, however, the deformation is relatively small, and the amount of spring back and plastic deformation are in a same range. If the amount of plastic deformation is overwhelming then the spring back can be neglected, and the object may be modeled as a rigid-plastic body. The spring cannot be considered as a rigid-plastic body. The objective of spring manufacturing process simulation is the estimation of spring back and the calculation of residual stress.

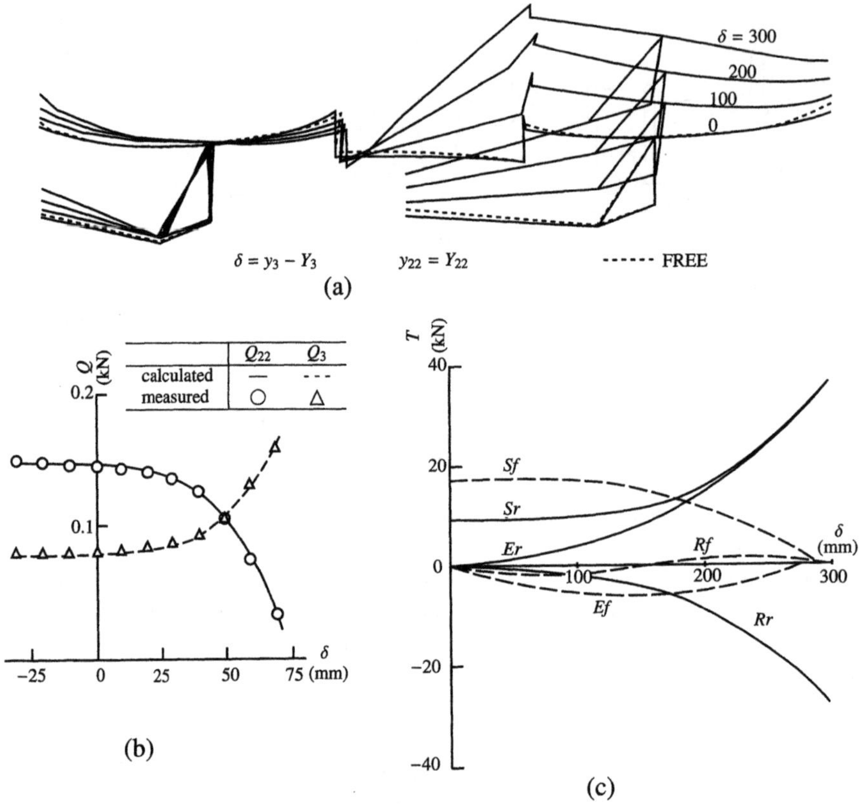


Figure 3.3: Analysis of the spring deformation when a car runs over a bump.

Therefore, it is indispensable to draw up the constitutive equations for the behavior of the transient segment from elastic to plastic region.

Having said this, the demand for the utilization of FEM information for the verification of manufacturing conditions has been intense. This takes us to the concept of CAE or computer aided engineering. If the calculation of presetting[9] or residual stress for helical spring, and/or the analysis of tightening stress for multiple leaf spring are available, the need for trial and error at the manufacturing site will be decreased favorably. Related to this, the simulation of the coiling process will be discussed in subsection 3.4.2.

Moreover, there are springs whose spring characteristics cannot be obtained without conducting the simulation of the manufacturing process. A typical example of such a spring is contact type spiral spring. The discussion will be introduced in subsection 3.4.2. Instead, initial tension will be described briefly here as an example of a manufacturing condition affecting the spring characteristics.

The initial tension is an interactive force between spring wires contacting each other. When a coil spring with initial tension is stretched, it does not show any deflection before the stretching force reaches the initial tension. Beyond that, the spring shows the same spring constant as a spring without the initial tension. In other words, the spring shows the characteristic biased by the amount of initial tension. In the case of torsion springs, this initial tension affects the characteristic of the spring as a friction factor. To incorporate this initial tension into helical spring, the spring shall be wound solid while restraining the torsion of wire, or more positively, giving a twist. This means that the magnitude of the initial tension is determined by manufacturing conditions.

Meaning, that the characteristic of helical spring with initial tension cannot be determined unless we carry out elasto-plastic analysis.

The implementation of such simulations is not as simple as might be considered only by introducing nonlinear analysis software. This is because a lot of data necessary for the analysis, such as the detailed mechanical property of wire in elasto-plastic domain and/or a friction coefficient between wire and tools, must be obtained through experiments.

## 3.4 PROSPECT OF FUTURE APPLICATION

### 3.4.1 Optimum Design

The "optimum design" will be accomplished by determining an objective function and the restricting conditions. The restricting condition can be expressed in two mathematical statement of equality and inequality. In the case of simplified formulas (3.1), and (3.2) for the cylindrical spring, for instance, the condition will be:

Equality: spring constant $k$ = specified value,
stress $\tau$ = design stress,
Inequality: outer diameter of coil < space limit,
installed height < space limit.

Our problem is to seek the optimum spring shape while satisfying the restricting conditions. In this case, there are three design variables: $d$, $D$, and $n$.

Although the object of optimization is diversified in each case, limit analysis is a common practice like pursuing minimum weight. The function that is calculated for the ultimate value, for instance a formula to express the weight in this case, is called the objective function. Within a category of simplified formulas, there are no significant problems in the optimum design process.

In the case of FEM, however, it is a different story. The first reason is that FEM is not suitable for the inverse problem. FEM is not a design tool but an analyzing tool as is described in section 3.1. The second reason is that the meaning of "optimum" becomes vague with the improvement of analyzing accuracy. When a cylindrical spring is compressed, the stress distribution along the wire shows a wavy pattern.

(The case study will be introduced later in section 5.3.) The conclusion leads us to a hypothesis that a piece of wire with wavy cross section along its length is an optimum shape to have uniform stress distribution. Although such a longitudinal contour of the wire may be ideal from the standpoint of minimum weight, it is an unrealistic solution. It is ironic that the selection of the objective function becomes more difficult when an improved analyzing technique becomes available. Research papers on the design of springs in which FEM is applied to the inverse problem have seldom been reported. However, a numerical analysis for the equalization of the surface principal stress on the cross section of thc wire was reported[10], and the method is already being applied to engine valve springs. Our future task will be to establish the way to link these optimum problems to FEM analysis.

### 3.4.2 Nonlinear Problems in Manufacturing Simulation

Two typical examples of nonlinearity concerning the spring manufacturing process will be introduced here. The first is the most popular spring manufacturing process of coiling. This process is very difficult, though it seems easy to handle because of its familiarity. The second example is a contact type spiral spring (commonly known as a power spring). This is tough example because the torque characteristic cannot be defined unless the manufacturing process is analyzed elasto-plastically, and the spring characteristic is nonlinear.

**(1) Simulation of Coiling** Even if a spring is to be formed to a specified coil diameter using the same material, the installation position or the angle of pin tools cannot be uniquely decided. There is no uniqueness in the solution. Nevertheless, this does not necessarily permit the inaccurate installation of tools. Even if the appearance of the spring is similar, there may exist such invisible differences as surface defects or the pattern of residual stress, which affect the quality of the product, e.g. fatigue life, sag resistance.

Delicacy in setting up coiling machine tools has been put into machine operator's hands thus far. It will be gradually shifted to the simulation technique. FEM is a reliable tool in this field. Some difficulty is anticipated along the way. The spatial configuration of the wire between the guide and the tool pin or even the length is unknown, so the object model is not solid enough for the discretization into elements. This problem must be treated as a moving-boundary condition problem. Further, the developed length of the wire should be affected by such manufacturing parameters like the feeding speed or the contact pressure to pitch tools. Before modeling, a careful preliminary examination by the factorial experiment is essential. Incidentally, a technical paper of chaos theory, advanced mathematics dealing with complicated systems, on the analysis of coiling conditions could be reffered to for this discussion[11].

**(2) Nonlinearity of Spiral Springs** The spiral spring is assembled in a case in which most of the flat wire is pressed against the inside wall of the case or outer layer of itself. When the shaft of the spring is turned, the stuck part of the spiral is separated in succession till all the length of the spiral is released. When the shaft is turned further, the inside portion of the spiral begins to stick around the shaft. Through a series of these moves, the spiral spring shows a torque characteristic shown in Fig. 3.4.

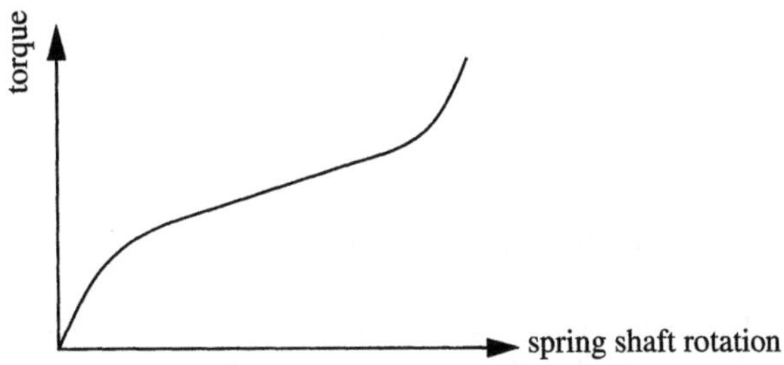


Figure 3.4: Characteristic of a spiral spring.

When the spring is wound, the torsional spring constant alters its degree along with the spring shaft rotation as follows;

gradual decrease → constant → gradual increase

When it's unwound, the curve does not trace the opposite direction of the winding curve since friction intervenes between coils.

When the spiral spring is analyzed by FEM, we may confront the following difficulties:

The degree of complexity of a spiral spring is higher than that of a multiple leaf spring, although they have the same contact problem. The contact problem of a spiral spring shall work on the continuous change of the contact point. In the case of a spiral spring, the relative slip between spirals is sizable and the contact partners are constantly changing for each increment of movement. The renewal of contacting partners requires a special treatment called rezoning. At the initial condition, after assembling the spiral in a case, the spiral is stuck against the inside wall of the case. This condition controls the loosening move of the spirals and affects the torque characteristic of the spring. This means that the analysis cannot be completed without knowing the initial stress at the installed condition. To determine the initial stress, the shape of the spiral in a free state after removing it from the case must be known. To learn the shape of the spiral in a free state, an elasto-plastic analysis in the forming process is essential. Thus a simulation of the working process is indispensable in the design analysis. A spiral spring normally goes through the following processes.

slitting → working of side-edges → annealing

→ cutting to length and forming of cut ends → primary winding
→ tempering → secondary winding (pre-setting)

Among the above processes, at least four steps must be taken into consideration: annealing, primary winding, tempering, and secondary winding.

In studying these aspects, it could be said that a spiral spring is very hard to solve by FEM. As a matter of fact, all of the required considerations can be implemented when we adopt a method of numerical solutions based on integral equations[12]. So, the necessity of FEM for the analysis of a spiral spring is debatable. In any case, the analysis of a spiral spring is the hardest problem among various springs from the standpoint of FEM.

### 3.4.3 Necessity of Material Data

We cannot evaluate the strength of springs even if the generated stress is obtained, unless we know the allowable stress of the spring material.

In the case of a helical spring, for instance, it is common to evaluate the fatigue life by the modified stress of Wahl (3.2). As a result, fatigue test data for the helical spring has been compiled from Wahl's modified stress. In FEM, any kind of stress can be easily obtained on request: principal stress, principal shear or equivalent stress. But the modified stress, which is physically vague in meaning, cannot be obtained at all by FEM. In any case, the analysis and the data shall work together to achieve the design task. If the analysis goes ahead leaving the data behind, it is nothing worthwhile. In design, FEM is at the mercy of the available data. Analysis is readily possible for making overall changeovers immediately after the introduction of FEM software, however, things are quite different where data is concerned. Almost all the data stored in the past will be discarded because of changes in the evaluation criteria. It will require quite a long time to compile new data. The change of the data cannot be accomplished in one day.

On the other hand, FEM brings us some convenience: it is liberation from the 'spring technology' that is specific to the spring in a small word. Changing to the universal FEM, we can use the data or the evaluating methods cultivated in fields other than spring. The 'spring technology,' is going to be incorporated into the general theory by using FEM as its breakthrough. The collapse of barriers dividing specific fields is to be welcomed from the standpoint of the comprehension and integration of the technologies. To this end, the total revaluation of conventional technology which have been developed independently within each specific field is inevitable.

## Bibliography of Chapter 3

[1] Watari and Kobayashi, General Equation for Design of Coiled Springs (in Japanese), Transactions of JSSR (1959), p89

[2] Takata M, Proc. 7th Japan National Congress Applied Mechanics (1957), p365

[3] Shimoseki et al, Analysis of Compression Spring by Matrix Method (in Japanese), Transactions of JSM A45 (1979), p901
[4] Shimoseki and Fujinuma, An Isoparametric Element for Curved and Twisted Beams (in Japanese), Transactions of JSM A55 (1989), p1809
[5] Shimoseki M, Machine Design Vol. 33 No. 11 (1989), p105
[6] Uruno et al, Geometrical Non linear Analysis for Helical Spring (in Japanese), Technical Report of JSSR Meeting (1996 Fall), p41
[7] Shimoseki and Hayasaka, RBSM Application for Non linear Problems (in Japanese), Technical Report of JSSR Meeting (1982 Fall), p47
[8] Kawai and Kondo, FEM Analysis of Buckling (in Japanese), Journal of SNAJ Vol. 142 (1977), p197
[9] Shimoseki and Sato, Inference of Change of Height due to Presetting in Coil Springs with varied Pitch and Coil Diameter (in Japanese),Transactions of JSSR (1979), p46
[10] Ohkuchi et al, Shape Optimization of the Wire Cross Section of a Helical Spring (in Japanese), Transactions of JSM A55 (1989), p2033
[11] Stewart I, WIRE INDUSTRY Sep. (1995), p491
[12] Shimoseki et al, On the Numerical Analysis of Spiral Springs (in Japanese), Transactions of JSSR (1988), p34

# 4 CLASSIFICATION AND APPLICATION OF ELEMENT

## 4.1 INTRODUCTION OF VARIOUS ELEMENTS

In chapter 2, the structure of FEM, i.e., the derivation process of stiffness equation and element stiffness matrix, was explained. It might be needlessly time consuming, if we had to follow these processes every time we solve a problem. Fortunately, there is no need for it. Because the stiffness equation is, as will be understood from the derivation process in chapter 2, basically invariable whether it belongs to two-dimensional problem or to three-dimensional problem. FEM is formulated and programmed so as to be applicable to a variety of problems. Several kinds of element shapes are prepared as a standard for the variety of modeling which requires different geometry. The element stiffness matrix for each element is also predetermined as a standard. Our effort is limited to select — according to the kind of the problem or the geometry of the object — the appropriate element from a pool of prepared elements. If we cannot find the appropriate element from the standard, new elements must be developed. In this case, the stiffness matrix of the new element will be gained by following derivation process in chapter 2. It is admittedly rare that we face a problem that needs element derivation, because a fairly abundant number of elements are usually prepared for the general purpose FEM software. The situation is quite similar in the case of spring calculation by FEM, and it is important to select appropriate elements corresponding to the type of spring.

Various kinds of elements for general FEM have been generously introduced in books on the market[1]. In this chapter, the characteristic of elements mainly used in chapters 5 and 6 will be described focusing the subject on springs.

### 4.1.1 Beam Elements

This element is the fundamental element second to the truss element in chapter 2. An example of the beam element is shown in Fig. 4.1.

This element is expressed by connecting two or more nodes with line segments and the geometrical shape is a straight line or a curve. It has a dimension of length. It is called a one-dimensional element on this account. The cross-section of the element does not appear directly in a practical geometrical model; it appears in

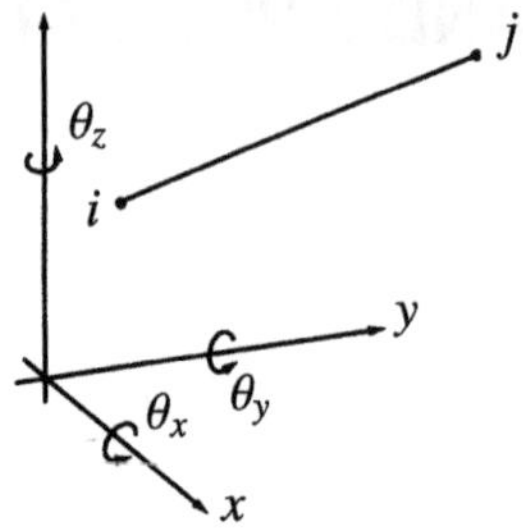


| Name | node | freedom | displacement function | integral point |
|---|---|---|---|---|
| 2 node 3-dimensional beam | 2 | 6 | linear/cubic | 1 |

(a)

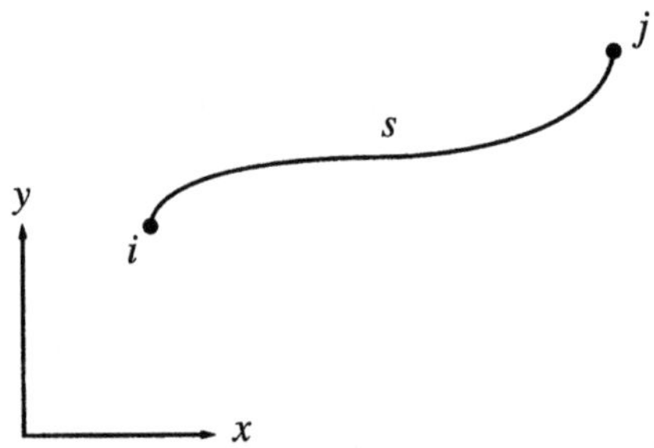


| Name | node | freedom | displacement function | integral point |
|---|---|---|---|---|
| 2-dimensional curved beam | 2 | 4 | cubic | 4 |

(b)

Figure 4.1: Example of beam element.

the element stiffness matrix as cross section area and the moment of inertia. This element is divided into two groups according to whether the nodes are connected with straight lines or curves. The degree of freedom of node differs according to whether it is on a plane or in space. There are three degrees of freedom in the case of a plane, and six in the case of space.

The concept of plane beam and space (or three-dimensional) beam elements is applicable to both straight beam and curved beam elements. Owing to the difference in the degree of freedom between the plane and the space, the number of rows in the element stiffness matrix also differs from each other.

**(1) Straight beam element** This element has two nodes and the nodes are connected by a straight line. There are two types of this element, that is, plane element with a node of three-degrees of freedom and space element with a node of six-degrees of freedom. The plane straight beam element deals with displacement within the plane including the element. These elements are straight beams and their applicable object is limited to straight bars. When applying to curved bars, an approximation by a polygon provides a limited usage of the straight beam element. To improve the accuracy of approximation, increased number of segments is necessary. Curvature effect cannot be taken into account even with improved accuracy.

**(2) Curved beam element** This element has two or more nodes and the nodes are connected by curves. Plane curved beam and space curved beam elements also exist depending on the degrees of freedom of the nodes.

There are two cases of plane curved beam elements. One is an element that deals with circular elements and the other is an element that deals with general curves. The former is called in-plane curved beam element and it deals with circular arcs. The latter is an element to deal with general quadratic curves.

The in-plane curved beam element has two nodes just like the straight beam element. It takes the effect of the curvature into account. The curvature is specified by the radius of the circular arc. In this event, the curvature between the nodes is constant as the region between the nodes is a circular arc.

The circular arc bar can be specified in a strict geometrical shape by applying the in-plane curved beam element. The in-plane curved beam element only treats circular arcs. It cannot be applied to a shape in which straight lines are included. Since the shape of a coil spring is a helix, it is necessary to take the torsion into account, as well as the curvature. Therefore the in-plane curved beam element is insufficient for the treatment of the helical spring.

The element that deals with general curves other than a circular arc is a beam element of higher order with three or more nodes. The higher order beam element is called a three-node (second order) element or a four-node (third order) element, depending on the number of nodes. The member between nodes is shaped by a curve with the order. The curvature is calculated by the curve function and it isn't a constant value between nodes. It can be used for curved bars of any shape. In differing from the in-plane curved beam element, the higher order beam element can deal with a complex assembly of straight lines and curves.

The spatially curved beam element also has two categories like a circular arc element and a general curve element.

The generally known spatially curved beam element is a circular arc element that takes curvature into account but not torsion rate. Beside the existence of report[2] which covers the combination of curvature and torsion, general-purpose software equipped to deal with such an element is rare.

The higher order beam element in space that covers general curve is basically the same as the element in a plane. The number of degrees of freedom in a plane is different from the one in space. In the case of the higher order beam element, it is theoretically possible to take curvature and torsion rate into account, depending on the number of orders. However, this advantage is not universally utilized because the main object of the element is the precise modeling of geometrical shape.

When the straight beam element and the higher order beam element are compared, the number of segments could be smaller and yet the accuracy of analysis is better in the latter since the displacement is approximated by a higher order function.

### 4.1.2 Plate Elements

The plate element is an element to be used in modeling flat plates or curved plates. It is a two-dimensional element with three or more nodes. The geometrical shape of the element is basically triangular or quadrangular. In terms of thickness, it could be constant or a function of the nodal coordinates. The element is also classified into three categories, which are plane elements, bent plate elements, and shell elements. These are classified by mechanical condition whether it treats in-plane deformation or out-of-plane deformation or both. These elements will be explained as follows.

**(1) Plate element with in-plane deformation** When a load is applied along the plane of a flat plate, the plate is deformed within the plane and tensile or compressive stress is generated. The plate element is used to analyze such in-plane deformation. An example of the plate element is shown in Fig. 4.2.

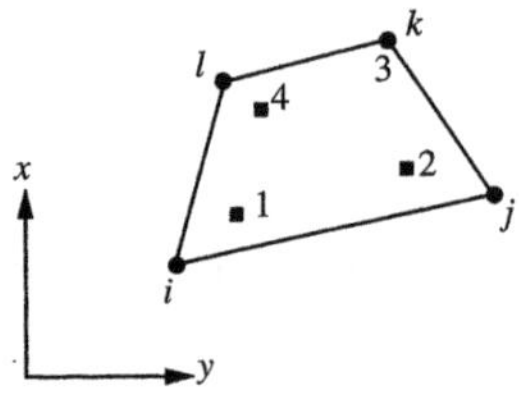


| Name | node | freedom | displacement function | integral point |
|---|---|---|---|---|
| Quadrilateral type | 4 | 2 | linear | 4 |

(a)

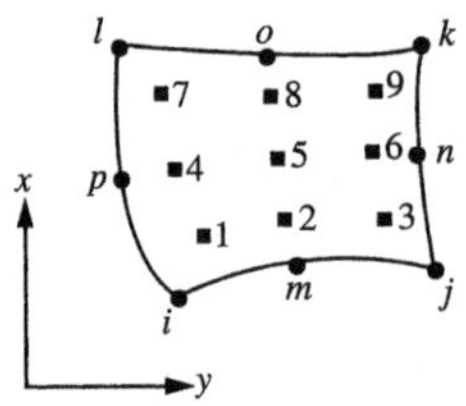


| Name | node | freedom | displacement function | integral point |
|---|---|---|---|---|
| Modified quadrilateral type | 8 | 2 | quadratic | 9 |

(b)

Figure 4.2: An example of plate element with in-plane deformation

There are several types of elements depending on the number of nodes or the boundary line condition like a straight or curved line.

**(2) Plate element with out-of-plane deformation** When a load or a bending moment is applied in the direction perpendicular to the plane, the plate deforms toward out of the plane direction and bending stress is induced in the plate. The plate element is used to analyze such out-of-plane deformation. An example of the plate element is shown in Fig. 4.3.

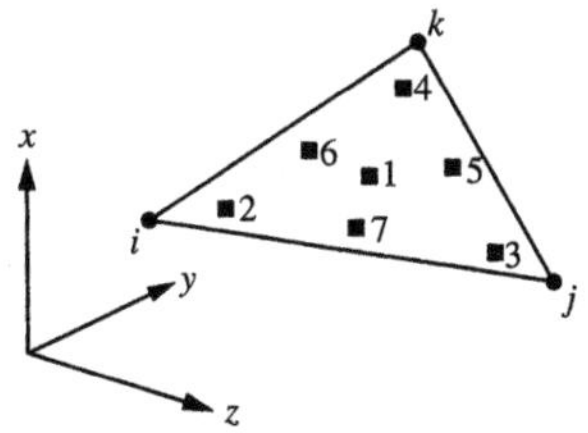


| Name | node | freedom | displacement function | integral point |
|---|---|---|---|---|
| triangular type | 3 | 5 | cubic | 7 |

(a)

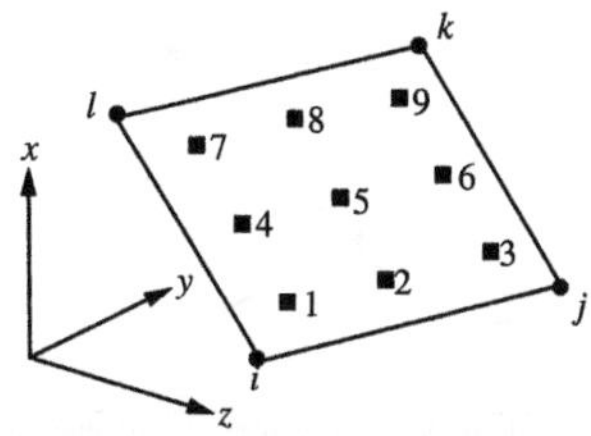


| Name | node | freedom | displacement function | integral point |
|---|---|---|---|---|
| Quadrilateral type | 4 | 6 | cubic | 9 |

(b)

Figure 4.3: An example of plate element with out-of-plane deformation.

The majority of the plate elements are based on Kirchhoff's plate theory. In-plane and out-of-plane deformation can be treated separately, if the thickness of the plate is sufficiently thin in this theory. According to this theory, the in-plane deformation or the plane stress problem can be treated by the element explained in (1), while the out-of-plane deformation or plate bending problem can be treated independently. It has sufficient accuracy of analysis for the flat and the thin plate, but cannot be applied to plates with a curved surface. When a curved plate is deformed to an out-of-plane direction, shell elements should be used. If the out-of-plane deformation of plates with a curved surface is to be treated by a plate element, the curved surface will be approximated by a polyhedron with a finer cut division. In that case, however, the plane stress within the plane can not be taken into account.

**(3) Shell elements** An example of the shell element is shown in Fig. 4.4.

There are two categories in the shell element, the plane shell element and the curved surface shell element. There is an essential difference between them. In the plane shell element, Kirchhoff's plate theory applies since in-plane deformation is independent of out-of-plane deformation. In the curved surface shell element, Kirchhoff's plate theory does not apply.

In other words, the plane shell element is an element developed on the basis of Kirchhoff's plate theory. The curved surface shell element is an element newly developed on the basis of a shell theory differing from Kirchhoff's plate theory.

Generally, there are two types of shell structures, the thin-walled and the thick-

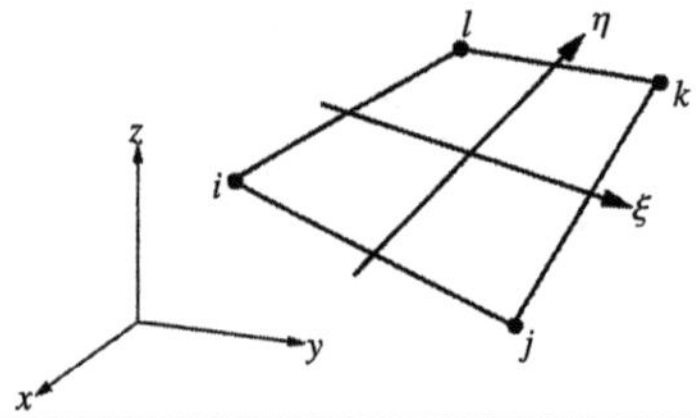


| Name | node | freedom | displacement function | integral point |
|---|---|---|---|---|
| Quadrilateral thick shell plane type | 4 | 6 | linear | 7 |

(a)

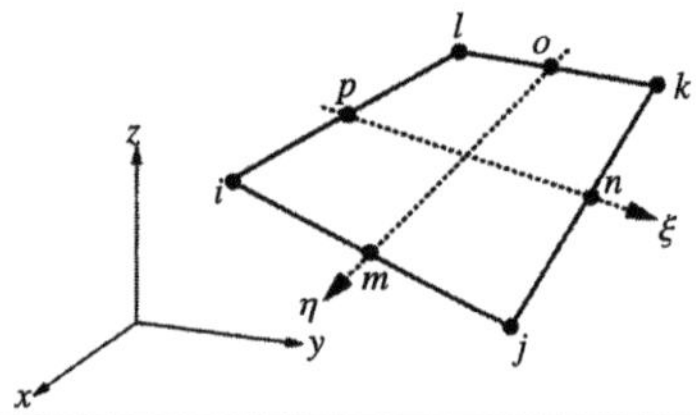


| Name | node | freedom | displacement function | integral point |
|---|---|---|---|---|
| 8 node thick shell curved surface type | 8 | 6 | quadratic | 7 |

(b)

Figure 4.4: An example of the shell element.

walled shell structure. Either of them are applicable to the curved surface shell element.

The isoparametric shell element is the element applicable to the thick-walled shell structure and stands on an assumption that the normal to the central plane of the isoparametric shell element remains straight after deformation and the stress in the normal direction is zero. This will be described in subsection 4.1.4.

### 4.1.3 Axisymmetric Elements

This element is used to analyze circular solids like a cylinder or a ring. An example of the axisymmetric element is shown in Fig. 4.5.

A circular solid is basically three-dimensional. However, the problem can be simplified by separating deformation in the rotating direction which is perpendicular to a virtual plane intersecting the axis and in-plane deformation on this plane. The axisymmetric element is applicable to the in-plane deformation of the circular solid. The axisymmetric element is divided into two categories, the axisymmetric body element and the axisymmetric shell element.

**(1) Axisymmetric body elements** This element is used to handle in-plane deformation on a cut plane which intersects the axis. It is applicable to parallel deformation but not to bending deformation. The element is geometrically a plane.

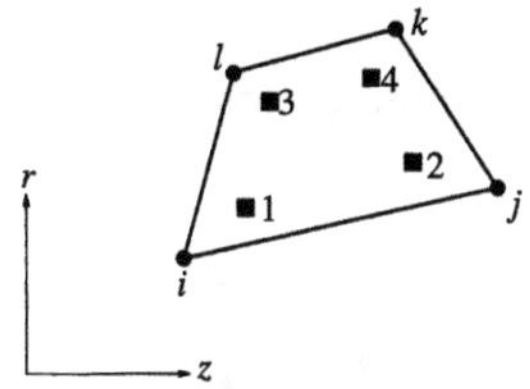


| Name | node | freedom | displacement function | integral point |
|---|---|---|---|---|
| Quadrilateral axisymmetric element | 4 | 2 | linear | 4 |

(a)

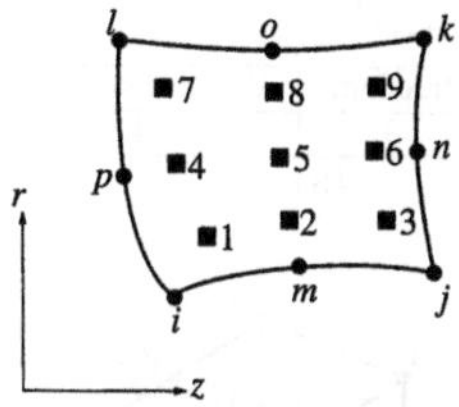


| Name | node | freedom | displacement function | integral point |
|---|---|---|---|---|
| Quadrilateral axisymmetric element | 4 | 2 | linear | 4 |

(b)

Figure 4.5: An example of the axisymmetric element.

In regard to the geometrical shape of the element, a triangle and quadrangle can be basic shapes. The element is divided into several groups depending on the number of nodes and whether the border of the element is a straight or curved line.

**(2) Axisymmetric shell elements** This element is used to handle in-plane deformation on a cut plane which intersects the axis in the axisymmetric shell structure. Differing from the axisymmetric body element, it is applicable to both parallel and bending deformation. An axisymmetric shell element is divided into two types. One is thin-walled type, and the other is thick-walled type.

The thin-walled axisymmetric shell element is a special case of the curved surface shell element and is used for the thin-walled axisymmetric shell structure. The element is divided into a straight line element and a curved line element depending on the geometrical shape of the principal axis.

The thick-walled axisymmetric shell element is a special case of the axisymmetric body element and is used for the thick-walled axisymmetric shell structure. The node number of the principal axis of the element ranges from two-nodal element, three-nodal element, four-nodal element, and so on. The geometrical shape of the element is linear in case of two nodes and a curved line in case of three or more nodes.

### 4.1.4 Cubic Elements (Solid Elements)

This element is used in modeling a three-dimensional body. An example of the solid element is shown in Fig. 4.6.

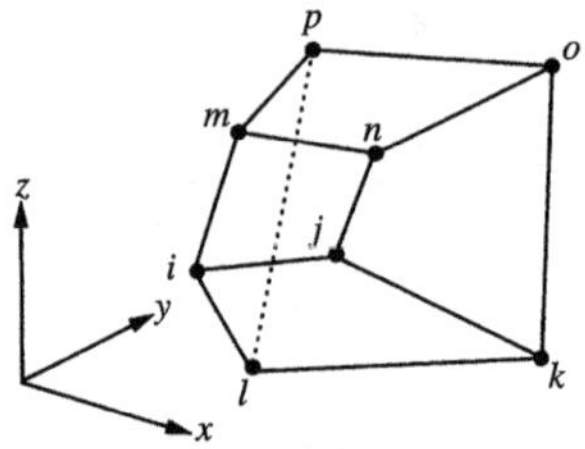


| Name | node | freedom | displacement function | integral point |
|---|---|---|---|---|
| 8 node solid element | 8 | 3 | linear | 8 |

(a)

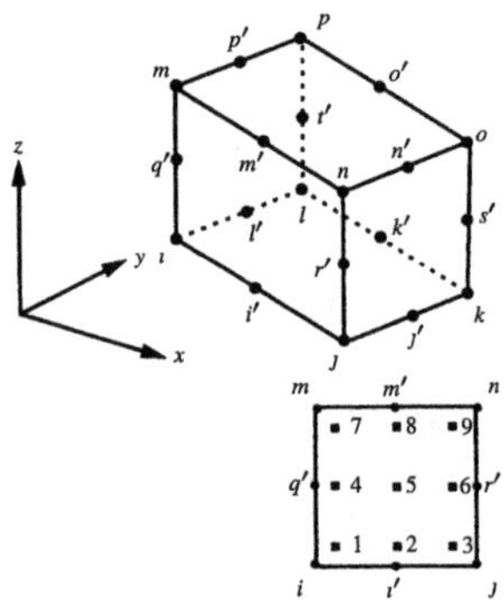

| Name | node | freedom | displacement function | integral point |
|---|---|---|---|---|
| 20 node solid element | 2 0 | 3 | quadratic | 274 |

(b)

Figure 4.6: An example of a solid element.

It is a three-dimensional element with four or more nodes. The basic geometric shape of the element is tetrahedron or hexahedron. It is also classified into simple polyhedron and isoparametric elements according to their shape function.

**(1) Simple polyhedron elements** This element is used mainly in modeling a three-dimensional body that does not include a curved surface. Among many polyhedrons, the tetrahedron and the rectangular solid (rectangular pillar) are mainly used as the elements for FEM. The nodes of the element are connected with straight lines and the surface of the element is a plane. The tetrahedron element has the fewest number of nodes and is the simplest element. This element constitutes a family of tetrahedron elements with a different number of nodes: for example, the 4-nodal first-order element, the 10-nodal second-order element, the 20-nodal third-order element, and so it continues. The displacement function for each element is expressed by an exact polynomial with three coordinate components for each node and the continuity condition of displacements between elements is always satisfied.

**(2) Isoparametric elements** This element is used mainly in modeling three-dimensional bodies with curved surfaces. This element is often called the 'distorted shape' element and is derived from the mapping method. In the case of mapping, the geometrical shape of the element is expressed by the shape function and the nodal coordinate, while the displacement within the element is expressed by the displacement function and the nodal displacement. There are several kinds of elements for the relation between the shape function and the displacement function; a typical element is the isoparametric element of which the shape function is common to the displacement function as a constituent function. Meaning that, the number of components of the nodal coordinate and the number of the components of the nodal displacement coordinate are equal. There are many kinds of these elements varying on the number of nodes. They can treat objects with very complicated geometry. With this element, the number of divisions is fewer than with the tetrahedron element, and better accuracy can be expected, in the analysis. The isoparametric element can also be applied to beam and plate element too. For specific details, refer to speciality books[2].

### 4.1.5 Contact Elements

A contact element is used in treating contact problems and is to be inserted between contact surfaces. This element was developed to provide an analytic model of the mechanical condition between contact surfaces; it is in a sense, a virtual element. A variety of contact elements have been developed by FEM software providers and are included in their general-purpose software packages. Each element has different characteristics depending on its developer. For the details of these elements, refer to their manuals. Only the outline of the contact element will be described here.

A contact element is classified into two groups according to contact condition we are considering. One is a GAP element in the case of contact between nodes, and the other is a joining element in the case of contact between elements.

**(1) GAP elements** This element is inserted between the nodes of contact surfaces; it originates from the beam element and has two nodes. This element is divided into two groups depending on whether friction is taken into account or not. One is constant-direction GAP element, and the other is constant-distance GAP element. The constant-direction GAP element accounts for the friction and the orientation of this frictional surface is predetermined by the coordinates of both nodes before deformation.

On the other hand, the constant-distance GAP element cannot take friction into account. Its simplest element originates from truss and adopts a method to define the stiffness of element as 'rigid' or 'zero' according to the condition of contact. These constant-direction or constant-distance GAP elements have the judgment equation of contact in their formulations. Among the elements that are inserted between contact surfaces, there is an element called a linear spring element. This element does not include the judgment of contact; it originates from beam element and is used for insertion between the nodes of contact surfaces. Therefore, this element is a kind of GAP element except for the absence of contact judgment. Because this element has an independent stiffness corresponding to the degree of freedom of nodes, the mechanically discontinuous surface could be handled as a virtually continuous surface by inserting this element, thereby enabling analysis.

**(2) Joining elements** The joining element is an element inserted between the elements of contact surfaces and has four or more nodes. Several kinds of elements are proposed according to the displacement function of longitudinal direction along contact surfaces. Some joining elements are used in analytical examples in Chapter 6. Refer to the operating manual of each general-purpose software package for details.

Although a GAP element is effective when the change of contact domain is small, a joining element is considerably more effective when the change of contact domain is large. Also, the joining element is more effective than the GAP element when contact domain is broad and contact position cannot be specified.

## 4.2 SELECTION OF ELEMENT AND DISCRETIZING PRACTICE

FEM is essentially an approximate solution. There are two meanings for this 'approximation': one is to approximate the geometrical shape of object by some appropriate elements and the other is to approximate displacement within element through displacement function. The approximation based on displacement function was explained in chapter 2. The subject in this section is the approximation of the geometrical shape of object. As long as the object is handled by an approximated model by elements, it is an important factor for the accuracy of analysis that how close the model is to the shape of the object. There may be a case where the model approximation is rigorous enough thanks to the closeness of element and object, but such cases are exceedingly rare.

In this section, the selection of element and the practice of discretization for the analysis of springs will be explained.

### 4.2.1 Selection of Elements

It is necessary to select elements from two distinct viewpoints: one is the geometry of the object and the other is the content of analysis. These are closely related to each other.

First, let us discuss geometry. The shape of the majority of springs is, in plain words, 'slender' or 'flat'. There are, of course, some exceptions. Formed wire springs, stabilizers, and helical springs are 'slender'. Multi-leaf springs and spiral springs are 'flat'. Some of the 'flat' springs could be classified as 'slender' from a different point of view. For instance, multi-leaf spring and spiral spring would belong to 'slender' shape based on a two-dimensional viewpoint. Most often, the beam element is suitable for 'slender' springs and the plate element is suitable for 'flat' springs. The axisymmetric element is appropriate to axisymmetric springs such as disc springs.

In addition, we have a cubic element to analyze three-dimensional bodies. While a beam element is a one-dimensional element with the factor of length and a plate element a two-dimensional element with the factor of area, the cubic element is a three-dimensional element with the factor of volume. In other words, the cubic element is the most practical element and basically all objects can be modeled by the cubic element in principle. All springs are three-dimensional bodies and they could be modeled based on the cubic element. As to be expected, during analysis by FEM, some disadvantages may arise in modeling all objects by a cubic element. These subjects will be explained in Paragraph 4.2.2.

Secondly, let us move to the subject of an analytical content. The analytical content varies from simple linear problems to complicated nonlinear problems. The element to be selected has a close relationship with the analytic contents.

In modeling the helical spring based on a beam element, for instance, a non-circular cross-sectional wire cannot be treated by the beam element, since the element cannot handle the cross section of the wire. In general, it is necessary to use the solid element to treat a non-circular cross-section.

In the analysis of a plate, torsion cannot be handled by the two-dimensional element. To treat the problem of torsion, it is necessary to carry out the modeling with an appropriate element. If the object has an axial symmetry, the number of dimensions of the element can be reduced by one.

As is mentioned above, it is necessary to take the shape of object and the content of analysis into account when selecting an element. Some typical examples based on the above viewpoint are listed below.

| | |
|---|---|
| Beam element: | Formed wire springs |
| | Helical springs |
| | Stabilizer bars |
| | Multi-leaf springs (two-dimensional) |
| | Spiral springs (two-dimensional) |
| Bent plate element: | Multi-leaf springs |
| | Spiral springs |
| Axisymmetric shell element: | Disc springs |
| | Ring springs |

For the examples in chapters 5 and 6, the analysis was carried out in this category except in a special case. The special case will be explained on each occasion.

### 4.2.2 Tips on Discretization

The next step after the selection of element and the determination of analyzing content is the discretization of domain into elements. The problem becomes, what part of the object should be discretized on what mesh size. In general, the domain with sharply changing physical quantities should be divided into fine elements. It has been proven that infinitely fine discretization converges into the exact solution. However, it is not advisable to divide the domain indiscriminately fine. Such dividing increases the number of elements, resulting in a lengthy and costly analysis. On the other hand, a rough discretization does not always guarantee satisfactory results. There are no general standards regarding the ideal fineness of discretization. The necessary fineness of the discretization varies on each case and there is no option but to rely on our own experience. The relation between the number of divisions and the accuracy of analysis will be described here based on an example of the linear analysis of a helical spring.

A cylindrical helical spring was selected as the object of analysis. The utilized element is a two-nodal three-dimensional straight beam element. For the boundary condition, the lower end of the spring was completely restrained, and the displacement of the upper end was limited to in the vertical direction. The variation of the nodal reactions based on a varying number of discretizations per coil was investigated and Fig. 4.7 shows the result.

It can be clearly seen that $w$ rapidly converges to a value obtained by rigorous analysis when the number of discretizations per coil comes to about 20. In the case

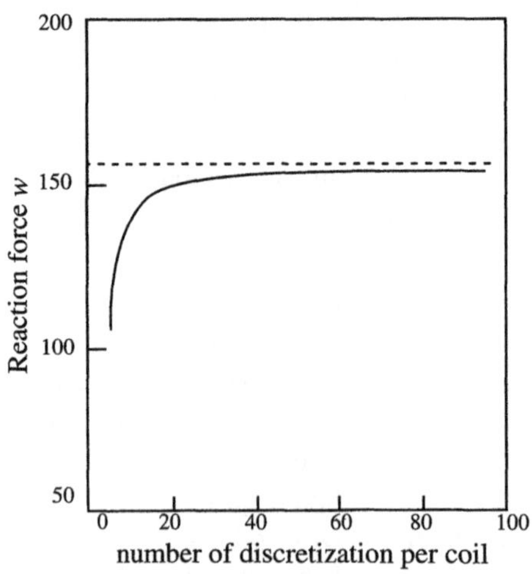


Figure 4.7: Analytical accuracy affected by number of discretizations (cylindrical spring).

of analysis for cylindrical springs by straight beam elements, adequate analytical accuracy can be obtained by 20 discretizations per coil. The broken line in Fig. 4.7 shows the analytical result for a cylindrical spring strictly modeled with a three-dimensional curved beam element.

In the above example of a cylindrical spring using beam elements, the shape of each element did not become a problem; the problem was the size of the element, since the beam element is one-dimensional and the shape of each element is a uniform segment. By way of comparison, in the case of two- or three-dimensional elements, such as a plate element or a solid element, not only the size but also the shape of each element affects the accuracy of solution. In the case of the triangular element, for instance, the shape shown in Fig. 4.8(b) should be avoided.

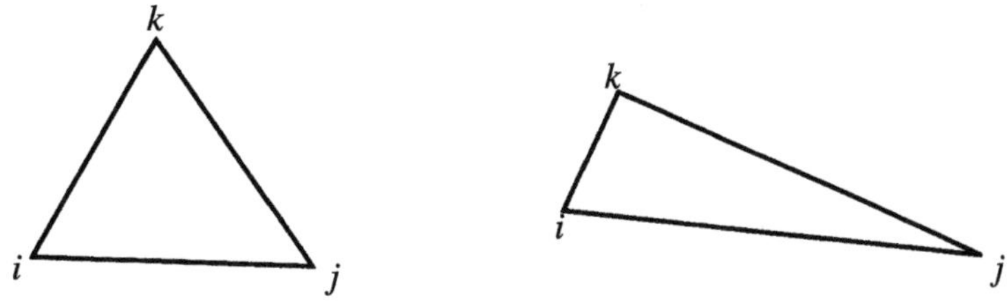

(a) (b)

Figure 4.8: Shape of divided elements.

It is ideal to discretize the domain with a shape shown in Fig. 4.8(a) as much as possible. The analytical accuracy of Fig. 4.8(b) is inferior to that of (a) due to the property of the element. Here are some tips.

As is explained in chapter 2, the displacement within the element is expressed by a displacement function. In the case of a triangular element, for instance, the displacement function in common use is a linear function. Meaning, that the displacement within the element is constant and the triangular element is a constant strain element or a constant stress element. If the stress gradient of the object is sufficiently small within a broad domain as is shown in Fig. 4.8(b), a comparably good approximation may be obtained by the element with such shape. However, when the stress gradient is relatively sharp compared to the domain size, the supposed to be constant stress in the domain does not hold. The figure shown in Fig. 4.8(a), which is close to an equilateral triangle, is more desirable than the figure in Fig. 4.8(b), which has a large aspect ratio in this regard. In terms of element size, a smaller one results in better approximation and higher accuracy of solution than does the larger one. This is why fine discretization is recommended for locations where sharp change in physical appearance takes place.

## Bibliography of Chapter 4

[1] Washizu et al, FEM Handbook (in Japanese) 1982, Baifukan

[2] Shimoseki M, An Isoparametric Element for Curved and Twisted Beam (in Japanese), Transactions of JSSR (1990), p25

# 5 ELEMENTARY ANALYSIS

The theme of this chapter is to let the weary designers, who want to use FEM in their spring design, have a chance to realize the easiness of FEM analysis.

The discussion will be limited to linear analysis here, and advanced analysis that takes geometrical nonlinearity into account will be discussed in chapter 6.

In the analysis of springs, it was intended to construct a model with basic elements such as beam elements, truss elements, shell elements, solid elements, and so on, avoiding special elements like contact elements or friction elements as much as possible. In the course of discussion, advice on the construction of FEM model and the analysis of solution will be explained. Furthermore, the analytical accuracy will be examined against the experimental results to confirm the limit of linear analysis.

The outline of this chapter is as follows. In section 5.1, the procedures and precautions for the preparation of input data for FEM analysis are explained on an example of a formed wire spring using a two-nodal three-dimensional beam element, which is one of the most fundamental elements. In section 5.2, an example of analysis for a stabilizer bar, one kind of formed wire spring, is introduced. In the case of hollow sections, it is necessary to assess flattening or thickness deviation in their cross-section due to bending or distortion at the bent section. It is difficult to carry out such an analysis by the application of beam elements, however, it might be an informative try to execute simplified analysis.

In sections 5.3 to 5.5, some analytical examples are given for helical spring category. Helical compression springs, helical extension springs, and helical torsion springs will be described. The effect of a pitch angle and an eccentric oblique load will be examined, particularly on helical compression springs, as was pointed out in chapter 3.

In section 5.6, an analytical example for spiral spring will be explained.

There are two types of spiral springs: contact type and non-contact type. The non-contact type of spiral spring is nonlinear, as is described in subsection 3.4.2. Non-contact spiral springs using four-nodal shell elements will be discussed.

Section 5.7 is an example of analysis for a multi-leaf spring. In this section, the analysis is carried out based on the concept of tip contact method, a conventional method for multi-leaf spring. The position of the contact is reexamined by comparing analytical result with experimental result. In this case, the spring leaf is modeled by a two-nodal beam element, and the contact condition between spring leaves is modeled by coupling the spring leaves with truss elements.

In section 5.8, the analysis of a suspension spring for the magnetic head of a hard disc drive is introduced as an example of flat springs. Four-nodal shell elements are used for the analysis.

In section 5.9, some analytical examples of stress concentration factor will be introduced; shell and the solid elements are used in the analysis.

For the sake of convenience, the stress concentration has been calculated by an ambiguous method of a safety factor in a conventional analysis.

FEM analysis can be an effective tool for the calculation of such a stress concentration factor.

## 5.1 FORMED WIRE SPRINGS

A spring that is made of circular or non-circular cross-sectional steel wire is generically called a formed wire spring. The variety of shape means that the design calculation should be carried out by various calculating formulas[1, 2, 3, 4] according to the shape of the spring. If we can find a calculating formula appropriate to the shape of the spring under design, we are fortunate. When a calculating formula corresponding to the shape of the spring is not at hand, we must create a calculating formula by a base theorem such as Castigliano's theorem. This formulation needs some knowledge about strength materials and calculus techniques. On the contrary, the analysis by FEM can be easily carried out without particular knowledge of strength materials. The necessary procedure is as simple as easy data input of shape, material, restraining condition, load condition. It is necessary, of course, to follow prescribed order and steps. FEM has more universality than mathematical formulas, since it can adjust itself to the change of design condition accompanying the change of shape. The universal FEM software may be rather hard to handle due to its high universality in some cases. A personal computer program for formed wire springs and stabilizer bars, in which the simplicity of operation is taken into account, has been developed[5] for practical use.

In this section, tips and precautions for data input will be explained in detail in the first step of FEM. The analysis will be carried out on a zig-zag spring.

**(1) Tips on modeling and input data** As is shown in Fig. 5.1, a zig-zag spring has a shape alternately combined of a half circle and a straight line. This spring is used for seat cushions.

There are spatial shape zig-zag springs and flat shape zig-zag springs according to the difference in its structure to be installed on a seat-frame. As a starter, a flat shape zig-zag spring with dimensions shown in Fig. 5.2 will be presented.

The selection of the element type and the discretization of the spring into segments is the first step. The simplest two-nodal straight beam element is selected here. This is to approximate the shape of Fig. 5.2 by joining a number of straight elements. In this case, the bent portion shall be finely divided so as to give a smooth shape of approximation as much as possible. The stress concentration as a curved beam cannot be calculated even if the approximation gets more accurate to the bent

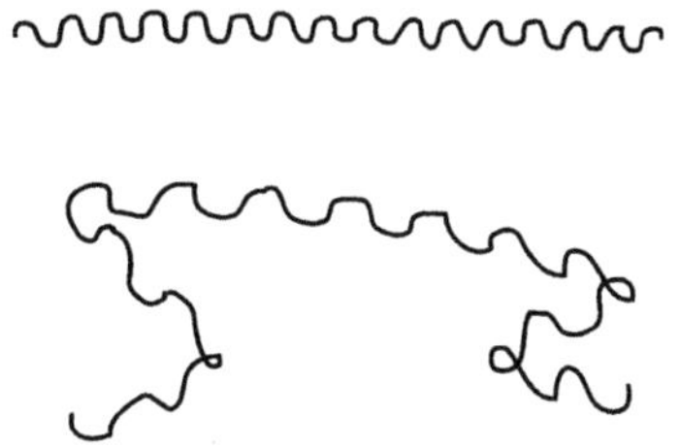

Figure 5.1: A zig-zag spring.

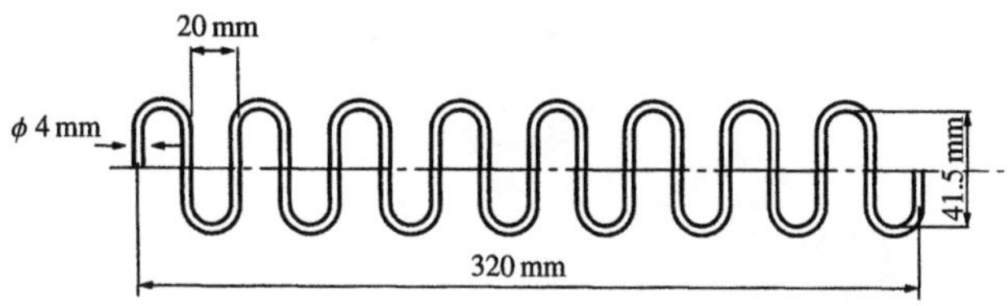


Figure 5.2: The shape and dimension of a zig-zag spring.

portion. The stress of this portion needs to be calculated separately based on bending moment and torsional moment obtained from FEM analysis. This will be mentioned later. Although the shape of the spring is two-dimensional, three-dimensional beam elements are selected here to be able to be applied to an out-of-plane load on the spring.

Generally, finer discretization is necessary in the area where the stress gradient is relatively large. The degree of fineness in discretization for a specific need depends on our experience. A closer solution will be obtained when the number of divisions is increased. However, a larger number of divisions requires a longer time for analysis. The discretization shall be carried out rationally, while balancing analytical accuracy with computer runtime.

A node shall always be placed on the spot where displacement is restrained or concentrated load is applied, since the boundary condition or the condition of concentrated load must be imposed on the node. Although the loading condition is imposed on the element (not on the node) when the distributed load is applied, the discretization shall be carried out taking the domain of the distributed load into account. Nodes and elements shall be numbered next. Although analytical efficiency may deteriorate slightly when the numbering for elements or nodes is not appropriate, recent FEM software packages have a faculty to renumber and thus, facilitate an effective analysis. This renumbering is a process within the computer and the result of the analysis will be indicated by the input number of node or element. The analytical model, taking all the above tips into account, is shown in Fig. 5.3. The numerals in the figure indicates the numbers of the nodes. The total number of nodes is 449 and the total number of elements is 448 in this model.

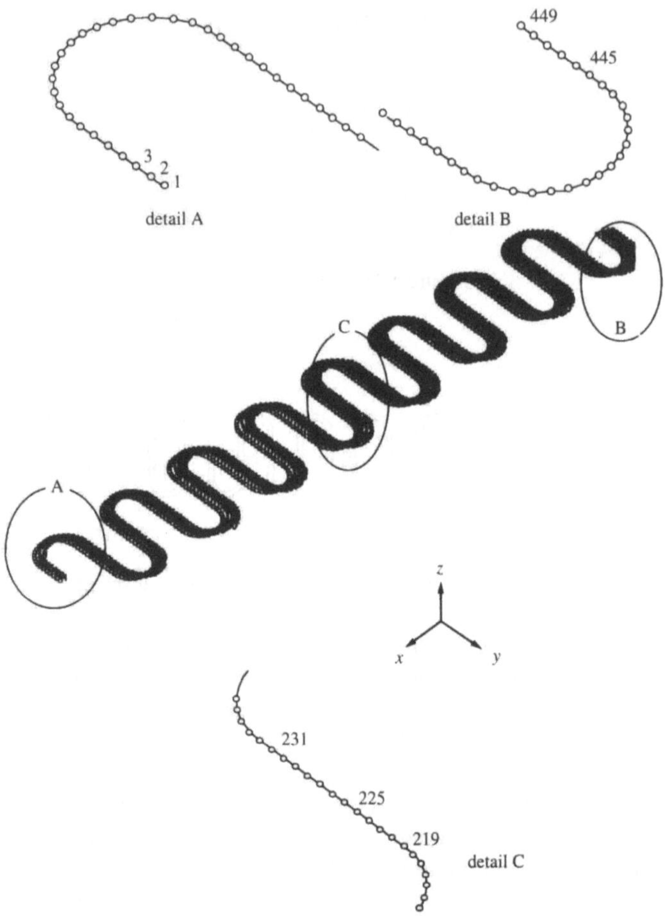


Figure 5.3: An analytical model.

The next step is the description of node and element data. The data to be prepared are the coordinates for each node, node numbers constituting each element as well as the type and cross-sectional characteristic of each beam element. Since doing figures on nodal coordinates in a bent portion is unexpectedly troublesome, it might be necessary to construct a calculation program or to use preprocessors or CADs on the market.

Table 5.1 shows nodal coordinates for two half-circles of the zigzag spring.

The element number and its node number are shown in Table 5.2. Table 5.3 shows the cross-sectional characteristic of the beam element.

For the type of beam element whose element stiffness matrix is obtained by integration, the necessary data is the shape of the cross-section and its dimension as

Table 5.1: Data of the nodal coordinates

| node NO. | $x$ (mm) | $y$ (mm) | $z$ (mm) | node NO. | $x$ (mm) | $y$ (mm) | $z$ (mm) |
|---|---|---|---|---|---|---|---|
| 1 | 320.00 | 0.00 | 0.00 | 29 | 300.00 | 0.00 | 0.00 |
| 2 | 320.00 | −2.10 | 0.00 | 30 | 300.00 | 2.10 | 0.00 |
| 3 | 320.00 | −4.19 | 0.00 | 31 | 300.00 | 4.19 | 0.00 |
| 4 | 320.00 | −6.29 | 0.00 | 32 | 300.00 | 6.29 | 0.00 |
| 5 | 320.00 | −8.38 | 0.00 | 33 | 300.00 | 8.38 | 0.00 |
| 6 | 320.00 | −10.48 | 0.00 | 34 | 300.00 | 10.48 | 0.00 |
| 7 | 320.00 | −12.58 | 0.00 | 35 | 300.00 | 12.58 | 0.00 |
| 8 | 319.83 | −14.39 | 0.00 | 36 | 299.83 | 14.39 | 0.00 |
| 9 | 319.30 | −16.26 | 0.00 | 37 | 299.30 | 16.26 | 0.00 |
| 10 | 318.37 | −18.05 | 0.00 | 38 | 298.37 | 18.05 | 0.00 |
| 11 | 317.07 | −19.65 | 0.00 | 39 | 297.70 | 19.65 | 0.00 |
| 12 | 315.47 | −20.24 | 0.00 | 40 | 295.47 | 20.24 | 0.00 |
| 13 | 313.68 | −21.87 | 0.00 | 41 | 293.68 | 21.87 | 0.00 |
| 14 | 311.82 | −22.41 | 0.00 | 42 | 291.82 | 22.41 | 0.00 |
| 15 | 310.00 | −22.58 | 0.00 | 43 | 290.00 | 22.58 | 0.00 |
| 16 | 308.18 | −22.41 | 0.00 | 44 | 288.18 | 22.41 | 0.00 |
| 17 | 306.32 | −21.87 | 0.00 | 45 | 286.32 | 21.87 | 0.00 |
| 18 | 304.53 | −20.94 | 0.00 | 46 | 284.53 | 20.94 | 0.00 |
| 19 | 302.93 | −19.65 | 0.00 | 47 | 282.93 | 19.65 | 0.00 |
| 20 | 301.63 | −18.05 | 0.00 | 48 | 281.63 | 18.05 | 0.00 |
| 21 | 300.70 | −16.26 | 0.00 | 49 | 280.70 | 16.26 | 0.00 |
| 22 | 300.17 | −14.39 | 0.00 | 50 | 280.17 | 14.39 | 0.00 |
| 23 | 300.00 | −12.58 | 0.00 | 51 | 280.00 | 12.58 | 0.00 |
| 24 | 300.00 | −10.48 | 0.00 | 52 | 280.00 | 10.48 | 0.00 |
| 25 | 300.00 | −8.38 | 0.00 | 53 | 280.00 | 8.38 | 0.00 |
| 26 | 300.00 | −6.29 | 0.00 | 54 | 280.00 | 6.29 | 0.00 |
| 27 | 300.00 | −4.19 | 0.00 | 55 | 280.00 | 4.19 | 0.00 |
| 28 | 300.00 | −2.10 | 0.00 | 56 | 280.00 | 2.10 | 0.00 |

Table 5.2: Constituting nodes of the element

| element NO. | node NO. | |
|---|---|---|
| 1 | 1 | 2 |
| 2 | 2 | 3 |
| 3 | 3 | 4 |
| 4 | 4 | 5 |
| ... | ... | ... |
| 444 | 444 | 445 |
| 445 | 445 | 446 |
| 446 | 446 | 447 |
| 447 | 447 | 448 |
| 448 | 448 | 449 |

Table 5.3: Cross-sectional characteristic of the element

| | |
|---|---|
| shape | round |
| radius (mm) | 2.00 |
| area, $A$ ($mm^2$) | 12.57 |
| moment of inertia of area (#1 axis), $I_{11}$ ($mm^4$) | 12.57 |
| moment of inertia of area (#2 axis), $I_{22}$ ($mm^4$) | 12.57 |
| the product of inertia of area, $J_{12}$ ($mm^4$) | 0.00 |
| the polar moment of inertia of area, $J$ ($mm^4$) | 25.13 |

a cross-sectional characteristic. When a round shape is chosen, the radius shall be specified. For the type of beam element whose element stiffness matrix is constructed with the analytical solution of the relation between load and displacement, the cross-sectional area, the moment of inertia of area, the product of inertia of area, and the polar moment of inertia of area shall be included as input data.

In the case of a beam element type whose element stiffness matrix is obtained by integration, the stress at the integral point of cross-section is numerated, and the stress calculation from cross-sectional force or cross-sectional moment is not necessary accordingly. The stress concentration as a curved beam is not factored into this case. It is convenient to use this type of beam element when the curvature at the bent portion is small with no stress concentration anticipated. When the stress concentration needs to be accounted for, we use beam elements where the element stiffness matrix is constructed based on an analytical solution. The latter type is used here. Then, the information expressing the relation between local coordinate system and global coordinate system ($x$-$y$-$z$ coordinate system shown in Fig. 5.4) is necessary. The positive direction of second axis of beam (local) coincides with the positive direction of $z$ axis (global) in this case study.

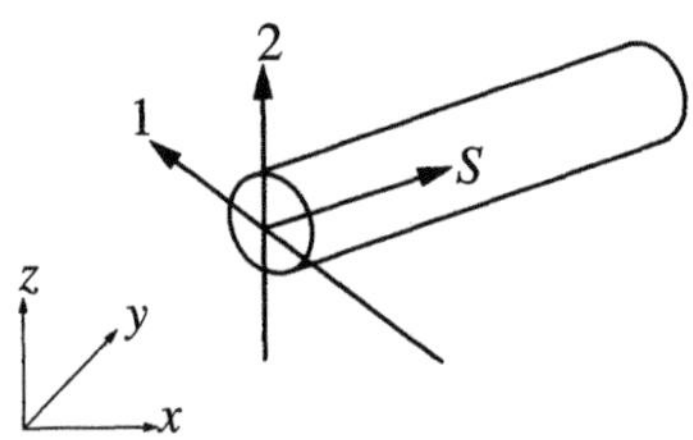


Figure 5.4: Coordinates of a beam element.

The next necessary information concerns material data. Since wire is manufactured by drawing process, it shows fibrous properties. For simplicity's sake, however, an isotropic material is assumed here. Since the analysis is restricted within the range of elasticity, only Young's modulus together with Poisson's ratio or alternately Young's modulus and shear modulus are required for the material data. Table 5.4 shows the material data used for the analysis.

The element stiffness matrix and the global stiffness matrix are automatically constructed in the computer based on the input data of Tables 5.1, ..., 5.4. As a

Table 5.4: Material characteristic

| | |
|---|---|
| Young's modulus, $E$ (GPa) | 205.8 |
| shear modulus, $G$ (GPa) | 78.4 |
| Poisson's ratio, $\nu$ | 0.315 |

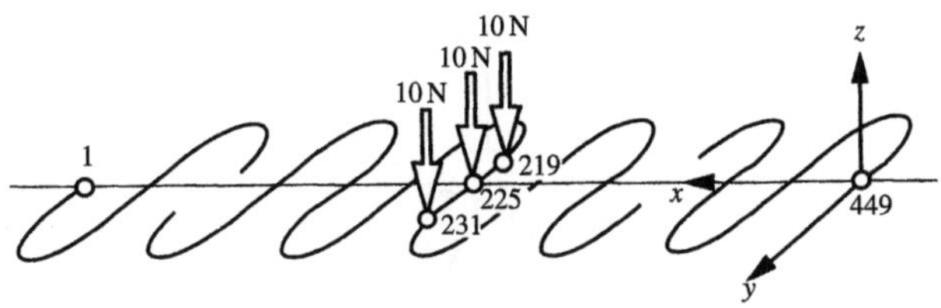


Figure 5.5: Boundary and loading condition.

final step, the input data of boundary condition and load condition shall be prepared considering the installing method or the usage of the spring. It is assumed here that the zigzag spring is installed on the seat frame so that both ends (nodes 1 and 449) of the spring can only rotate about $y$-axis of the spring. The normal load $P$ is applied to the center of the spring as shown in Fig. 5.5.

The input data of boundary condition and load condition is shown in Tables 5.5 and 5.6.

Table 5.5: Boundary condition

| node NO. | displacement | | | rotation | | |
|---|---|---|---|---|---|---|
| | $x$ | $y$ | $z$ | about $x$-axis | about $y$-axis | about $z$-axis |
| 1 | fix | fix | fix | fix | free | fix |
| 449 | fix | fix | fix | fix | free | fix |

Table 5.6: Load condition

| node NO. | $P_x$ (N) | $P_y$ (N) | $P_z$ (N) |
|---|---|---|---|
| 219 | 0.0 | 0.0 | −10.0 |
| 225 | 0.0 | 0.0 | −10.0 |
| 231 | 0.0 | 0.0 | −10.0 |

Our remaining task is to feed the data from Tables 5.1, ..., 5.6 to a computer, following the instructions of FEM analyzing software. Then, the analysis will be carried out automatically and the result will be obtained.

**(2) Postprocessing and analytical accuracy** The nodal displacement, axial force ($P_s$), transverse shearing forces ($P_1$ and $P_2$), bending moments ($M_1$ and $M_2$) and torsional moment $M_s$ are the result of the calculation (See Fig. 5.4 for notation). The stress is obtained by these cross-sectional forces and cross-sectional moments. At the bent portion, it is necessary to take the stress concentration into

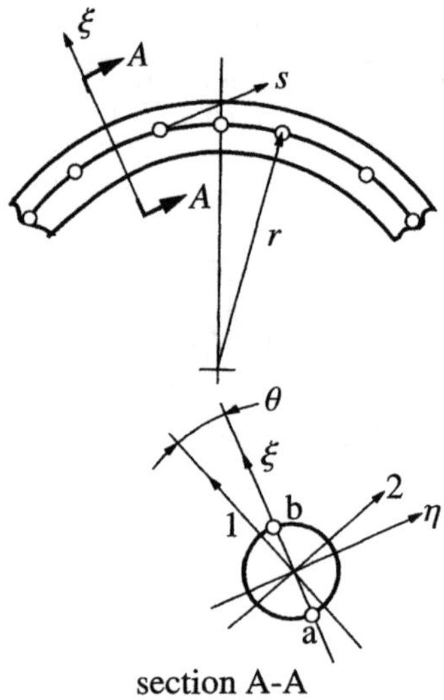


Figure 5.6: A local coordinate system.

account. As an example, the computational formula for the stress at points a and b in Fig. 5.6 are to be specified here.

First, we convert the output of cross-sectional force and cross-sectional moment, from the element coordinate system (See Fig.5.4), into $P_\xi$, $P_\eta$, $P_s$, $M_\xi$, $M_\eta$, and $M_s$, of the $\xi$-$\eta$-$s$ coordinate system shown in Fig.5.6 by equation (5.1). $\xi$, $\eta$, and $s$ indicate the principal normal, binormal, and tangent in the axial direction, respectively. $\theta$ is the angle formed by the primary axis of the element and $\xi$-axis.

$$\begin{Bmatrix} P_\xi \\ P_\eta \\ P_s \\ M_\xi \\ M_\eta \\ M_s \end{Bmatrix} = \begin{bmatrix} \cos\theta & \sin\theta & 0 & 0 & 0 & 0 \\ -\sin\theta & \cos\theta & 0 & 0 & 0 & 0 \\ 0 & 0 & 1 & 0 & 0 & 0 \\ 0 & 0 & 0 & \cos\theta & \sin\theta & 0 \\ 0 & 0 & 0 & -\sin\theta & \cos\theta & 0 \\ 0 & 0 & 0 & 0 & 0 & 1 \end{bmatrix} \begin{Bmatrix} p_1 \\ p_2 \\ p_s \\ M_1 \\ M_2 \\ M_s \end{Bmatrix}. \tag{5.1}$$

The normal stress $\sigma$ at point a is obtained by superposing the stress caused by $P_s$ and bending moment $M_\eta$. To obtain the stress by $M_\eta$, however, the effect of curved beam must be taken into account.

$$\sigma = \frac{4c^2 - c - 1}{4c(c-1)} \left( \frac{32 M_\eta}{\pi d^3} \right) + \frac{P_s}{A}. \tag{5.2}$$

Here, $d$ is the wire diameter and $c$ is the spring index ($= 2r/d$), in which $r$ is the radius of the curve.

The shearing stress $\tau$ at point a is obtained by superposing the stress caused by the shearing force $P_\eta$ and torsional moment $M_s$. A correction factor, proposed by Wahl, shall by applied to the stress by $M_s$.

$$\tau = \frac{(4c-1)}{(4c-4)}\left(\frac{16M_s}{\pi d^3}\right) + \frac{4.92P_\eta}{\pi d^2}. \tag{5.3}$$

In a similar way, the normal stress $\sigma$ and the shearing stress $\tau$ at point b, will be specified by (5.4) and (5.5), respectively.

$$\sigma = \frac{4c^2+c-1}{4c(c+1)}\left(\frac{32M_\eta}{\pi d^3}\right) + \frac{P_s}{A}, \tag{5.4}$$

$$\tau = \frac{(4c+1)}{(4c+4)}\left(\frac{16M_s}{\pi d^3}\right) - \frac{4.92P_\eta}{\pi d^2}. \tag{5.5}$$

The maximum principal stress $\sigma_{\max}$ and the maximum shearing stress $\tau_{\max}$ are to be obtained by the following formulas.

$$\sigma_{\max} = \frac{\sigma}{2} + \sqrt{\left(\frac{\sigma}{2}\right)^2 + \tau^2}, \tag{5.6}$$

$$\tau_{\max} = \sqrt{\left(\frac{\sigma}{2}\right)^2 + \tau^2}. \tag{5.7}$$

Fig. 5.7 shows the distribution of the transversal shearing force $P_2$.

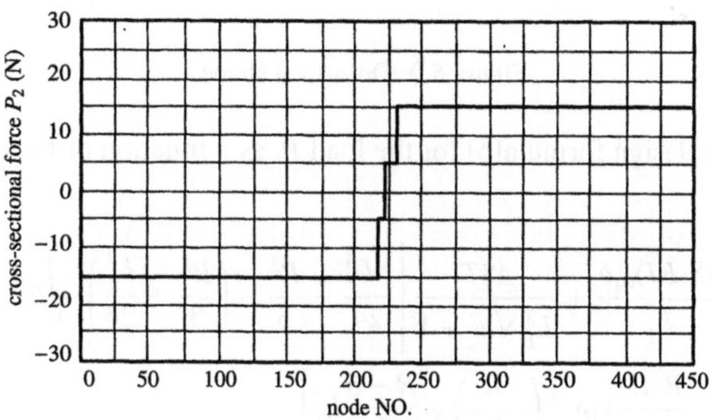


Figure 5.7: Distribution of cross-sectional forces.

Fig. 5.8 shows the distribution of the bending moment $M_1$ and the torsional moment $M_s$.

Fig. 5.9 shows the deformed shape of the spring based on the calculated nodal displacements.

Let us compare the result of FEM with that of the design formula concerning the relationship between the load $P$ and the displacement $\delta$ at the loading point. To

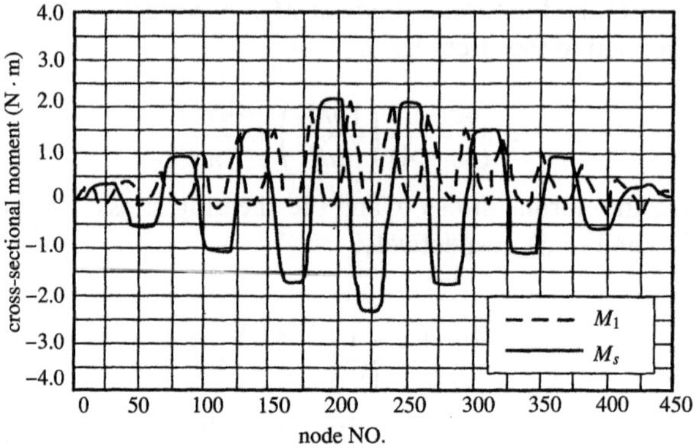


Figure 5.8: Distribution of the cross-sectional moment.

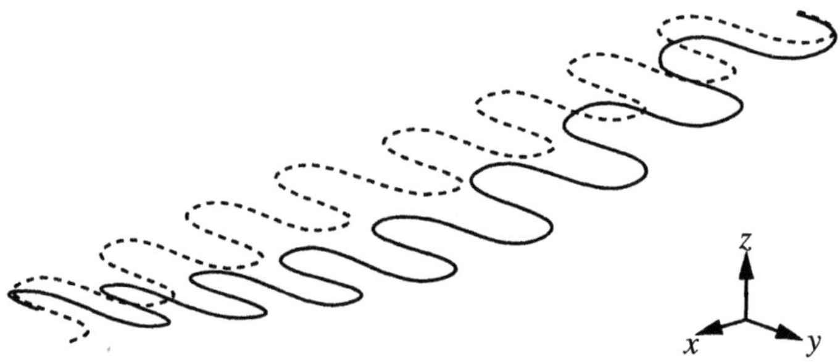


Figure 5.9: Deformed shape.

begin with, the design formula[6] for the load $P$, as a function of the deflection $\delta$, is as follows:

$$P = \frac{48(EI)_{\mathrm{eq}}\delta}{L_1^3} + \frac{48T}{L_1^3\sqrt{a^2+1}}\left[\frac{aL_1^3}{24} + \frac{R^3}{3} - \left(\frac{R^2}{3} + \frac{L_1^2}{24}\right)\sqrt{R^2 - \frac{L_1^2}{4}} + \frac{R^2L_1}{4}\sin^{-1}\left(-\frac{L_1}{2R}\right) + \frac{R-\delta}{8}L_1^2\right], \tag{5.8}$$

where $L_1$ is the installed length of the spring, $a\ (= L_1/2(R-\delta))$ is the inclination at the end point after deformation. This formula was derived based on two assumptions. One is that the zigzag spring is assumed to be a beam with an equivalent flexural rigidity $(EI)_{\mathrm{eq}}$. The other one is that the distribution of the deflection takes the form of the quadratic curve $y = \sqrt{R^2 - (x - L_1/2)^2} - (R - \delta)$ with a uniform radius $R$ as is shown in Fig. 5.10 and the internal tension $T$ is induced within the spring.

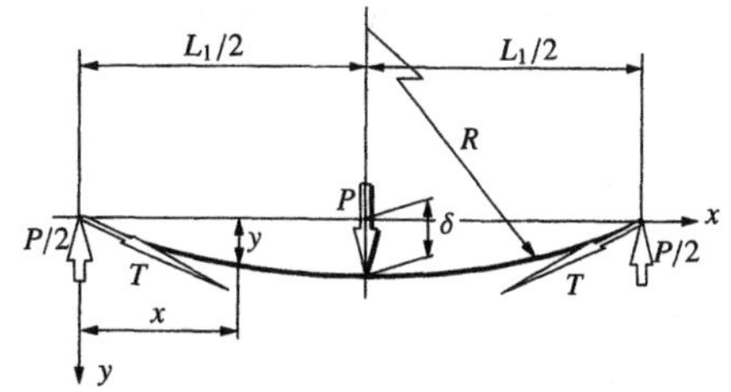


Figure 5.10: Assumption in the conventional calculating formula.

The internal tension $T$ is given by $k(S - S_0)$, where $S_0$ is the initial axial length of the spring, $S$ is the axial length of the spring after deformation ( $= 2R\cos^{-1}(1 - \delta/R)$), and $k$ is the tensile spring constant. The equivalent flexural rigidity $(EI)_{eq}$ and the tensile spring constant shall be calculated separately in advance.

The relation between load and deflection calculated by FEM and (5.8) is shown in Fig. 5.11 for comparison.

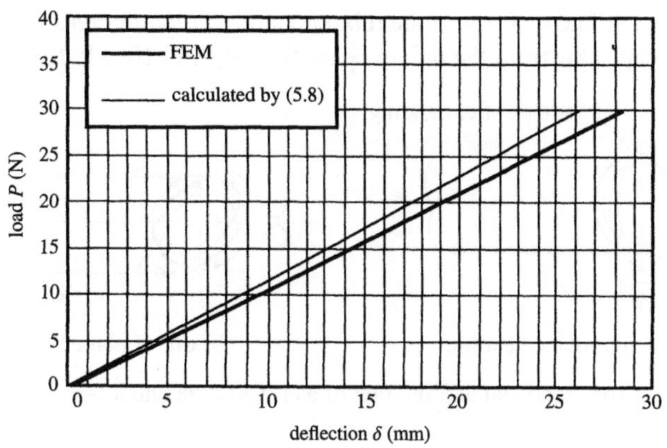


Figure 5.11: Load deflection characteristic.

As is clearly seen from the figure, the calculated result by the conventional design formula shows about 8% less deflection than the FEM analysis. This difference occurs because the distribution of the deflection was assumed to be a quadratic curve with a uniform radius $R$ in the conventional design formula. The deformation of the spring was restricted by this assumption.

Although the zigzag spring was analyzed under a concentrated load in the above analysis, the distributed load might be more appropriate when the spring is used as a seat spring. Also, the installing structure for the spring may differ from this demonstration analysis. In the conventional method by the design formula, the derivation of new formulas is often necessary when the load condition or the installing condition (boundary condition) has changed. In FEM analysis, however, there is flexibility with a new situation by just changing load condition and boundary condition.

## 5.2 STABILIZER BARS

A stabilizer bar is an auxiliary spring used to generate spring action corresponding to the stroke difference of left and right wheels reducing the rolling motion of car body for better steering characteristics. The configuration of the stabilizer bar takes a variety of complicated three-dimensional shapes owing to its design concept that makes it run through various peripheral parts underneath a chassis. Therefore, no universal design formula has been developed thus far. The design depends on a rough method of using formula for a simplified shape[7] with a trial and error method.

The majority of cross-sectional shapes used for a stabilizer bar are solid circles, and yet a hollow circular section has become popular in recent years. This hollow section helps to pursue weight reduction. When a bending moment is applied to the bent portion of a thin walled tubular stabilizer bar, the cross section flattens as shown in Fig. 5.12. The stiffness of the tubular stabilizer bar decreases and the stress distribution deviates far from the one in beam theory[8, 9, 10, 11, 12].

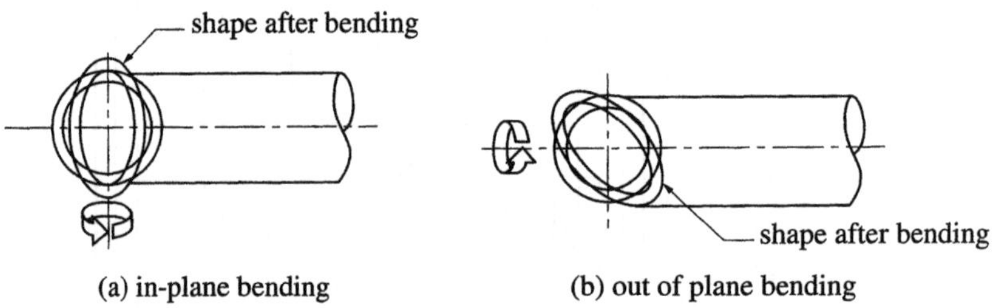


Figure 5.12: Cross-sectional deformation of a curved tube under bending moment.

The flattening of cross-section shall be considered to accurately obtain spring constant and stress. It is also necessary to consider that the initial shape of cross-section at the bent portion is not circular as is shown in Fig. 5.13, since the bending process might cause a deviation in thickness or the flattening of cross-section.

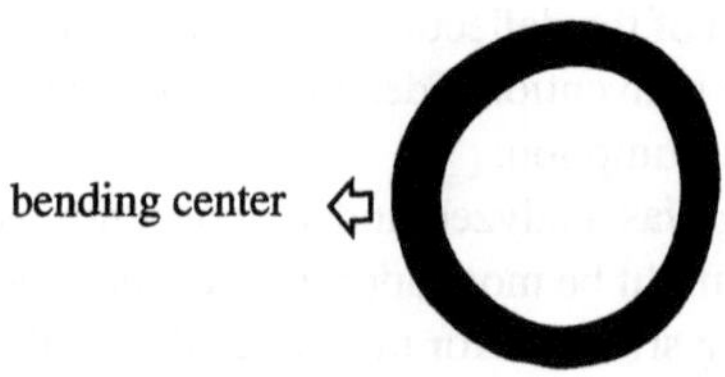


Figure 5.13: Cross-sectional shape at the bent portion.

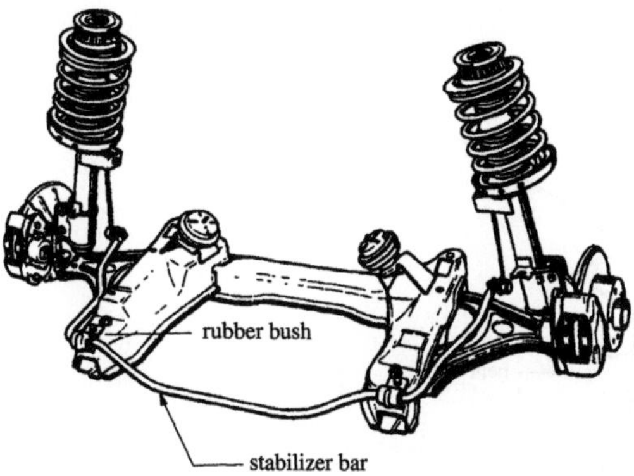


Figure 5.14: An example of a stabilizer bar installed in a chassis.

Furthermore, the stabilizer bar is generally installed on a car body using a rubber bush as is shown in Fig. 5.14. The rubber bush acts as a kind of spring.

The effect of the bush cannot be ignored when evaluating the rolling stiffness of vehicles and the stress generated in the stabilizer bar accurately.

To undertake an analysis taking these effects into account, shell or spring elements shall be used accordingly. However, the simple beam element is used here in the sample analysis. The analysis that takes the effect of the clamp bush or the flattening of the cross section into account will be discussed in chapter 6. The case of a tubular stabilizer bar will be presented here.

**(1) Points for modeling and input data** The analytical model is shown in Fig. 5.15.

The numerals in the figure are node numbers. The total number of elements is 112 and the total number of nodes is 113. The stabilizer bar is modeled by a pipe element, which is a kind of three-dimensional two-nodal beam element. The points of consideration or precaution in the discretization are similar to those in the case of formed wire spring in section 5.1.

The nodal coordinate value, the cross-sectional characteristics and the material characteristics of the beam element of this analytical model are shown in Table 5.7, Table 5.8, and Table 5.9, respectively.

Since the stabilizer bar is symmetrical with respect to *y*-*z* plane at node number 57, only the nodal coordinate values from No.1 to No.57 are listed in Table 5.7. The outer radius and the thickness of the cross-section are selected as the characteristics of the element, since the stiffness matrix is constructed by integration.

No.$n$ node and No.$n+1$ node constitute No.$n$ element. The second local axis of the cross-section is selected in parallel to the global *x*-*y* plane as shown in Fig. 5.16.

In Fig. 5.16, the direction of *s* axis shall be oriented to the node point increment direction.

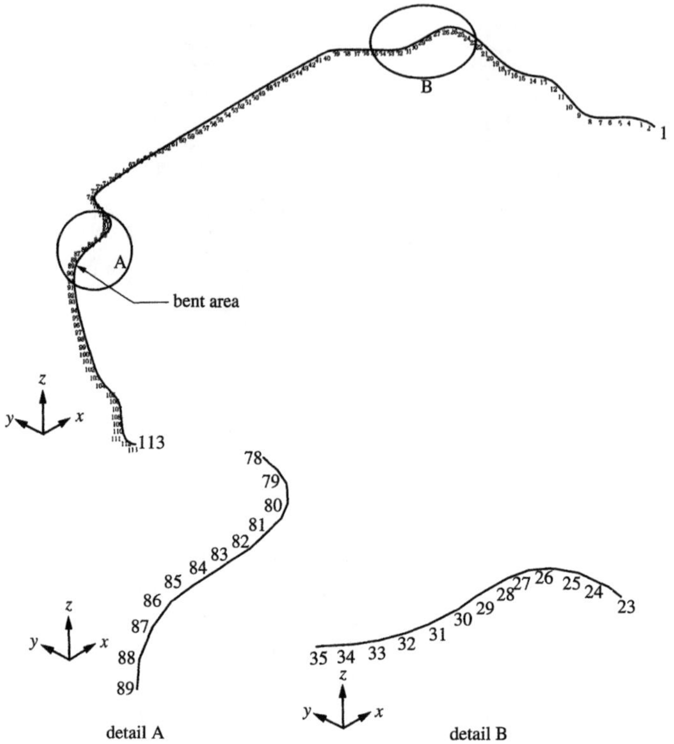


Figure 5.15: An analytical model.

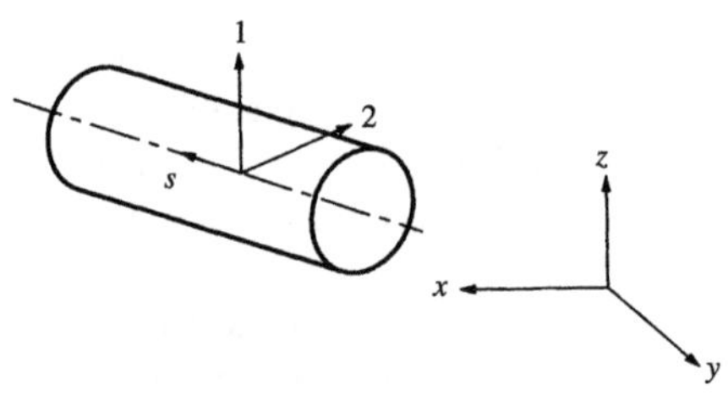


Figure 5.16: The element and global coordinate system.

Table 5.10 shows the boundary condition of displacement.

At the position of the rubber bush, the stabilizer bar is allowed to rotate about its own axis and forced displacement is applied to each end of the arm in the opposite directions.

The displacement of node No.30 in all directions is restrained to obtain a fixed rigid body. Nodes No.28, 30, 84 and 86 are nodes at the edge of the bush and nodes No.1 and 113 are tip points.

Table 5.7: Nodal coordinate values

| node NO. | $x$ | $y$ | $z$ | node NO. | $x$ | $y$ | $z$ |
|---|---|---|---|---|---|---|---|
| 1 | 553.00 | −276.00 | −129.80 | 30 | 343.19 | 0.00 | 0.00 |
| 2 | 553.00 | −262.15 | −129.80 | 31 | 329.67 | 1.51 | -0.33 |
| 3 | 551.03 | −248.25 | −129.80 | 32 | 316.33 | 6.20 | −1.37 |
| 4 | 545.12 | −235.21 | −129.80 | 33 | 304.33 | 13.95 | −3.09 |
| 5 | 535.97 | −224.56 | −129.80 | 34 | 298.68 | 24.11 | −5.34 |
| 6 | 526.20 | −215.99 | −129.80 | 35 | 287.97 | 35.67 | −7.90 |
| 7 | 516.42 | −207.42 | −129.80 | 36 | 281.12 | 51.21 | −11.34 |
| 8 | 505.59 | −195.28 | −129.80 | 37 | 274.45 | 66.74 | −14.78 |
| 9 | 498.40 | −181.07 | −122.59 | 38 | 267.69 | 82.28 | −18.22 |
| 10 | 496.00 | −167.08 | −114.45 | 39 | 260.31 | 94.17 | -20.85 |
| 11 | 496.00 | −157.37 | −107.80 | 40 | 249.26 | 103.90 | -23.01 |
| 12 | 496.00 | −147.65 | −101.14 | 41 | 235.77 | 110.02 | -24.36 |
| 13 | 493.29 | −132.50 | -92.59 | 42 | 221.67 | 112.00 | -24.80 |
| 14 | 485.48 | −117.53 | -87.57 | 43 | 206.89 | 112.00 | -24.80 |
| 15 | 476.50 | −104.83 | -84.91 | 44 | 192.11 | 112.00 | -24.80 |
| 16 | 467.95 | -92.74 | -82.36 | 45 | 177.34 | 112.00 | -24.80 |
| 17 | 460.70 | -82.01 | -78.29 | 46 | 162.56 | 112.00 | -24.80 |
| 18 | 454.78 | -72.25 | -70.92 | 47 | 147.78 | 112.00 | -24.80 |
| 19 | 449.94 | -63.68 | -62.51 | 48 | 133.00 | 112.00 | -24.80 |
| 20 | 448.89 | -54.72 | -53.71 | 49 | 118.22 | 112.00 | -24.80 |
| 21 | 439.66 | -45.46 | -44.62 | 50 | 103.45 | 112.00 | -24.80 |
| 22 | 434.40 | -36.14 | -35.47 | 51 | 88.67 | 112.00 | -24.80 |
| 23 | 429.14 | -26.82 | -26.33 | 52 | 73.89 | 112.00 | -24.80 |
| 24 | 422.54 | −18.16 | −17.83 | 53 | 59.11 | 112.00 | -24.80 |
| 25 | 412.87 | −10.52 | −10.33 | 54 | 44.33 | 112.00 | -24.80 |
| 26 | 400.75 | -4.68 | -4.59 | 55 | 29.56 | 112.00 | -24.80 |
| 27 | 387.21 | −1.14 | −1.11 | 56 | 14.78 | 112.00 | -24.80 |
| 28 | 373.48 | 0.00 | 0.00 | 57 | 0.00 | 112.00 | -24.80 |
| 29 | 358.34 | 0.00 | 0.00 | | | | |

Table 5.8: Cross-sectional characteristics of the element

| cut shape | doughnut |
|---|---|
| outer diameter (mm) | 11.1 |
| wall thickness (mm) | 3.2 |

Table 5.9: Material characteristics

| | |
|---|---|
| Young's modulus, $E$ (GPa) | 205.8 GPa |
| shear modulus, $G$ (GPa) | 78.25 GPa |
| Poisson's ratio, $\nu$ | 0.315 |

Table 5.10: Boundary condition of displacement

| node NO. | $x$ | $y$ | $z$ |
|---|---|---|---|
| 1 | free | free | 60.0 |
| 113 | free | free | –60.0 |
| 28 | free | 0.0 | 0.0 |
| 30 | 0.0 | free | 0.0 |
| 84 | free | 0.0 | 0.0 |
| 86 | frec | 0.0 | 0.0 |

**(2) Analytical result and accuracy** Initially, let us examine the analytical accuracy of stress. The stress is computed from cross-sectional force and cross-sectional moment. The distribution of cross-sectional force and cross-sectional moment are shown in Fig. 5.17 and Fig. 5.18, respectively.

In these figures, the abscissa shows the numbers of the elements and the ordinate shows cross-sectional force or cross-sectional moment. $P_S$, $P_1$ and $P_2$ are axial forces in each direction designated in Fig. 5.16. $M_1$, $M_2$, and $M_s$ are bending moments about each axis.

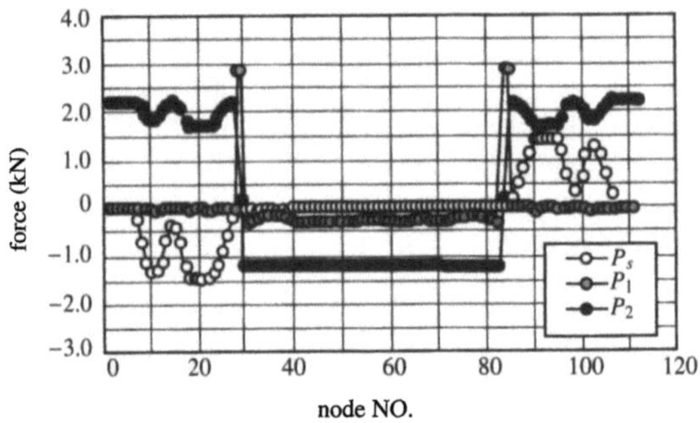


Figure 5.17: Distribution of the cross-sectional force.

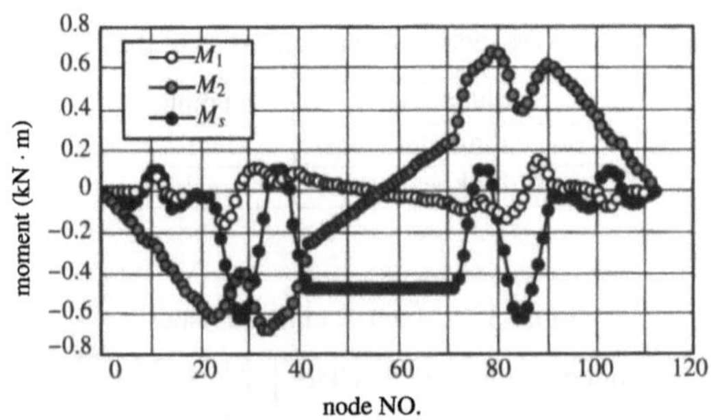


Figure 5.18: Distribution of the cross-sectional moment.

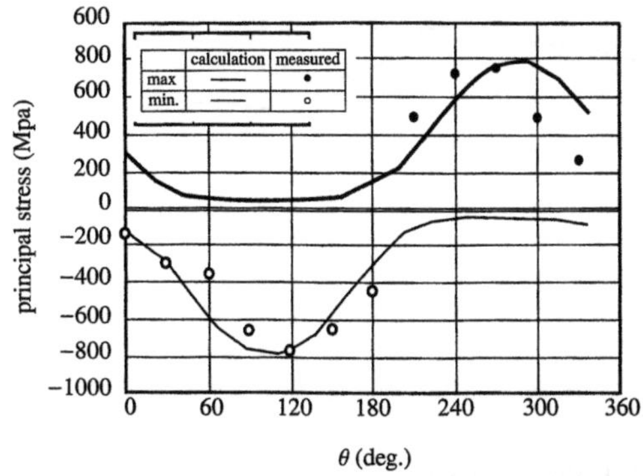


Figure 5.19: Distribution of the principal stress (curved portion in area A).

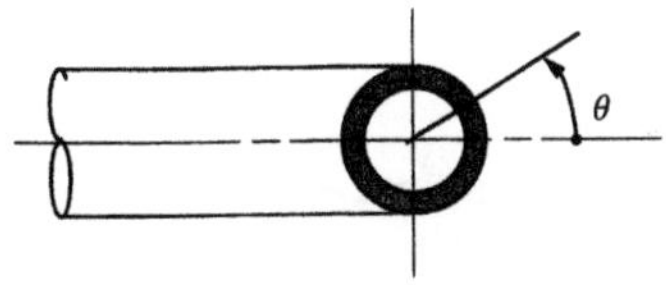


Figure 5.20: Angular position.

Fig. 5.19 is a result of the cross-sectional stress distribution at the bent portion A in Fig. 5.15, calculated from cross-sectional force and moment in comparison with measured value. (Stress concentration factor is not multiplied.)

The abscissa of the figure is angular position $\theta$ in Fig. 5.20, and the ordinate is principal stress. The solid lines are the analytical results.

Dots (solid and hollow) in the figure show the experimental results in case that rubber bushs are used. The analytical result of stress seems compatible with the experimental one by comparison, the former being only about 5% higher than the latter. The position of the maximum or minimum of the stress shows some difference between analytical and experimental results. This disagreement can be attributed, as explained before, to the convenient analysis that ignored the deformation of the cross-section induced by applied load or moment.

Next, let us see the load vs. deflection characteristic. The comparison between analysis and experiment is shown in Fig. 5.21.

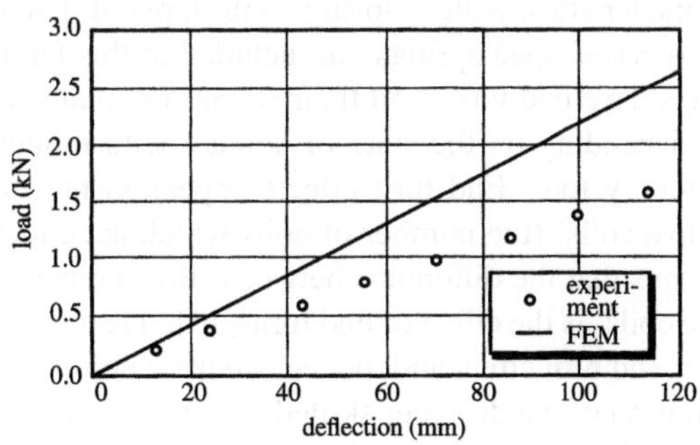


Figure 5.21: Load deflection characteristic.

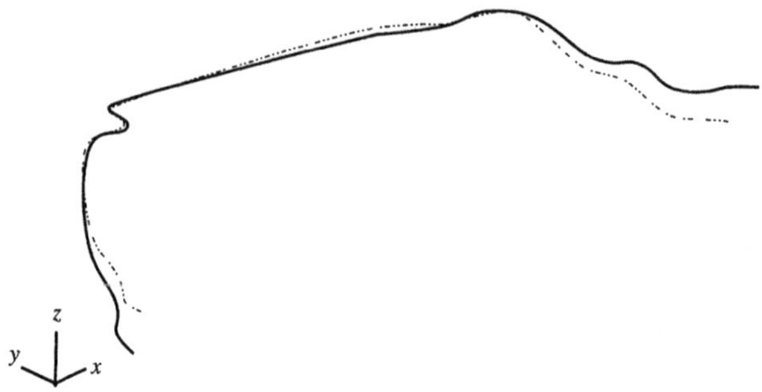


Figure 5.22: Deformed shape.

The experimental results show the case when rubber bushs are used. The abscissa of the figure is the relative displacement in $z$ direction between both arm ends and the ordinate is the applied load. The analytical spring constant 21.8 kN/mm is higher than the experimental 14.1 kN/mm by a 55% margin. The neglection of the rubber bush effect and the cross-sectional deformation seems to be the reason for the difference. This clearly demonstrates that the deformation of cross-section cannot be neglected for tubular stabilizer bar analysis. Fig. 5.22 shows the deformed shape of a stabilizer bar in this sample analysis.

In this figure, a solid line shows the shape after the deformation and the dotted line shows the shape before the deformation.

## 5.3 HELICAL COMPRESSION SPRINGS

The most commonly used spring in the world is the family of helical compression spring. A large variety of springs, namely, cylindrical springs, conical springs with nonlinear load characteristics, helical springs with tapered material, barrel-shaped spiral springs, hourglass-shaped springs are included in this family. All the helical compression springs have end turns and their shapes are many and various. These end turns deform depending on the state of contact with mount ring surface and they store strain energy too. End turn effects appear in the industrial jargon of 'the number of active coils' (the number of coils which act effectively as a spring) implicitly. This means that the difference between 'the number of active coils' and 'the number of free coils' is the effect of end turns[13]. There's not a lot of literature on the evaluation of end turn effect and, hence, we have no option but to rely on the experience and intuition of the designer skilled in conventional designing practice.

As was mentioned in subsection 3.1.1, many assumptions are associated with conventional design formula, including spring constant and stress, namely, (1) negligence of the effect of a spatially curved beam, (2) the external force will act on the central axis of the coil, (3) the deformation is minute. Naturally enough, calculated results may deviate from what we anticipate as the spring itself deviates from these assumptions. For springs with a small number of free coils and a large pitch angle, in particular, phenomena that are vastly different from conventional behavior are to be seen. For example, the stress on the outer diameter of coils is higher than that of inner diameter, and the distribution of stress is not uniform, and so on. To accurately assess phenomena like this, it is necessary to consider the effect of pitch angle, eccentric oblique load, and the curvature or the torsion of coils when constructing a calculation model.

A shape factor like a pitch angle is a purely geometrical problem and is easily analyzed by expressing the shape with beam elements. However, spatially curved beam element[14] which takes the curvature and torsion into account is not equipped in commercially available FEM software, and it is necessary to compute a new stress from the cross-sectional force and the cross-sectional moment taking the curvature and the torsion into account[15].

The rest of the problem is the effect of eccentric oblique load. When a spring is compressed, force transmits from mount ring to end turn. Whether the effect of eccentric oblique load can be evaluated accurately or not depends on how exactly the state of transmission of this force is expressed.

The modeling of contact between mount ring and end turn is the first important consideration in the FEM analysis of the helical compression spring. Then in the case of the spring with nonlinear load characteristic, contact between coils takes place when compressed. An adequate way to construct the model of this contact condition is another important point. The characteristic and stress of a helical spring of various shapes could be accurately analyzed using FEM analysis if these points are kept in mind.

### 5.3.1 Static Analysis

A simple analysis will be presented here excluding the condition of end contact and the geometrical nonlinearity.

**(1) Points in modeling and input data.** For the spring in this presentation, shape and specification are shown in Fig. 5.23 and Table 5.11.

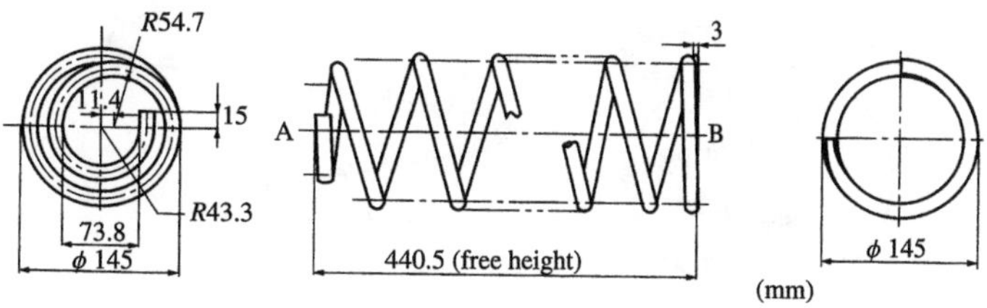


Figure 5.23: Shape and dimension of analyzed helical spring

Table 5.11: Specification of analyzed spring

| | | |
|---|---|---|
| wire dia. | | $\phi$ 12.8 mm |
| material | | SUP7 (JIS G 4801) |
| A | end turn shape | pigtail |
| | number of end turn | 0.75 |
| B | end turn shape | closed taper |
| | number of end turn | 1.00 |
| total coils | | 7.35 |
| spring constant | | 18.6 N/mm |

Let us start from points and precautions in the case of a helical spring constructed with a three-dimensional beam element model. As was described in chapter 4, there are various kinds of three-dimensional beam elements. By selecting an element with a higher order of displacement functions, we can analyze the problem with higher accuracy and in a short time, even if the number of divisions is small. In the case where contact between spring end and mount ring is to be analyzed, it is necessary to increase the number of nodes.

The simplest two-node uniform cross-sectional beam element is selected here and the number of discretizations will be 20 per coil. The analytical error of displacement is estimated to be less than 2.5% based on the diagram shown in chapter 4 (Figure 4.7).

Fig. 5.24 is the discretizing scheme of the analyzed helical spring.

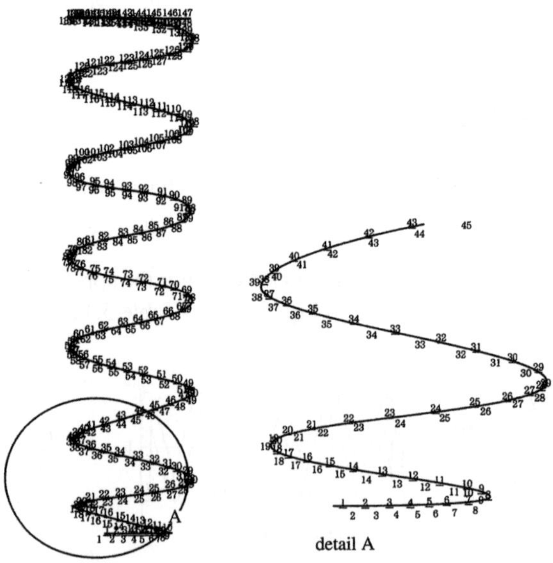


Figure 5.24: Discretization of helical compression spring.

Table 5.12: Nodal coordinates

| node NO. | $R$ (mm) | $\theta$ (deg.) | $z$ (mm) |
|---|---|---|---|
| 1 | 42.92 | 0.00 | 0.00 |
| 2 | 43.36 | 18.00 | 0.19 |
| 3 | 43.58 | 36.00 | 0.45 |
| 4 | 43.62 | 54.00 | 0.80 |
| 5 | 43.56 | 72.00 | 1.26 |
| 6 | 43.50 | 90.00 | 1.85 |
| 7 | 43.52 | 108.00 | 2.58 |
| 8 | 43.52 | 126.00 | 3.48 |
| 9 | 43.36 | 144.00 | 4.57 |
| 10 | 43.24 | 162.00 | 5.84 |
| 11 | 43.73 | 180.00 | 7.20 |
| 12 | 45.20 | 198.00 | 8.60 |
| 13 | 47.31 | 216.00 | 10.09 |
| 14 | 49.55 | 234.00 | 11.74 |
| 15 | 51.72 | 252.00 | 13.60 |
| 16 | 53.87 | 270.00 | 15.64 |
| 17 | 56.08 | 288.00 | 17.86 |
| 18 | 58.42 | 306.00 | 20.20 |
| 19 | 60.95 | 324.00 | 22.61 |
| 20 | 63.40 | 342.00 | 25.05 |
| 21 | 65.20 | 0.00 | 27.44 |
| ... | ... | ... | ... |
| ... | ... | ... | ... |
| 139 | 66.41 | 324.00 | 437.78 |
| 140 | 66.41 | 324.00 | 437.98 |
| 141 | 66.35 | 0.00 | 438.00 |
| 142 | 66.23 | 18.00 | 437.76 |
| 143 | 66.06 | 36.00 | 437.30 |
| 144 | 65.92 | 54.00 | 436.65 |
| 145 | 65.82 | 72.00 | 435.99 |
| 146 | 65.89 | 90.00 | 435.61 |
| 147 | 66.25 | 108.00 | 435.79 |
| 148 | 66.97 | 126.00 | 436.76 |

Plain figures are node numbers and underlined figures are element numbers in Fig. 5.24. The magnification of the lower end coil is shown in detail A. The number of elements in the model is 147 and the number of nodes is 148. The nodal coordinates are listed in Table 5.12 in a cylindrical coordinate system $(R, \theta, z)$ shown in Fig. 5.25.

The element data of this model is listed in Table 5.13.

Radius and the side length of a and b shown in Fig. 5.26 for the rectangular cross section are listed in the table.

The end turn of the closed taper spring is modeled as an aggregate of beam ele-

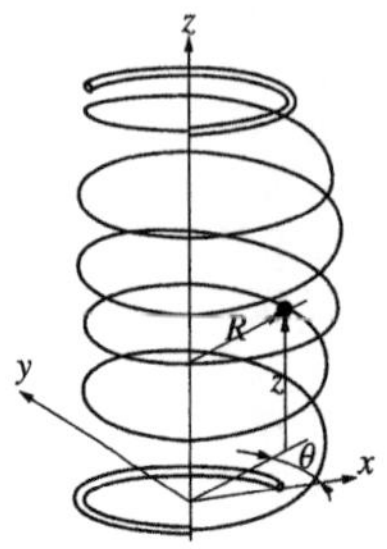


Figure 5.25: Cylindrical coordinate system $(R, \theta, z)$.

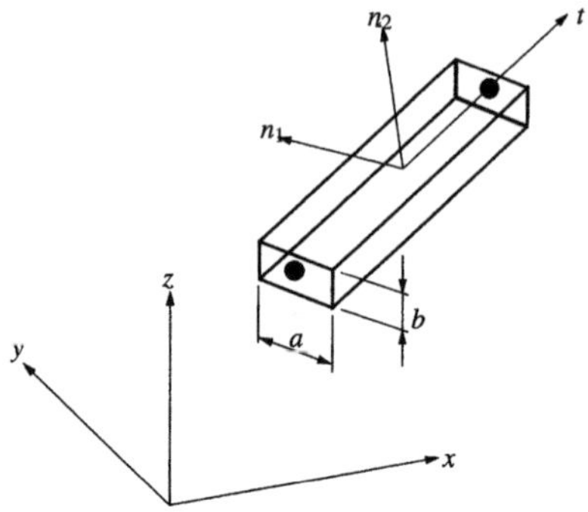


Figure 5.26: Element coordinate system $(t, n_1, n_2)$ of a rectangular beam element.

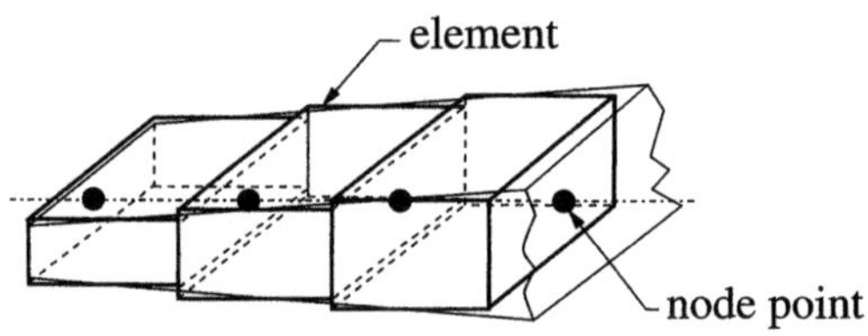


Figure 5.27: Model of taper portion (closed taper end).

ments with uniform cross section as shown in Fig. 5.27 and cross-sectional dimensions in Table 5.13 designate the value at the middle point of each element.

Then the data that shows the relation between the element coordinate and the global coordinate is necessary; the direction data of the basic vector ($n_1$ or $n_2$) of the element coordinate shown in Fig. 5.26, for instance. (Data table is not available.)

In Fig. 5.26, the notation $t$ means tangential vector of beam axis and the incremental direction of the node number is taken as positive. The data regarding material characteristics is shown in Table 5.14.

The global stiffness of the helical spring will be derived from the data in Tables 5.12, ..., 5.14 and the relation between element coordinate and global coordinate of each element.

Now let's get back to subsection 1.3.2 regarding the analytical procedure of FEM. To solve the equation, it is necessary to set boundary condition following the calculation of global stiffness. Normally, the helical compression spring is used under a parallel compression load and, therefore, we will discuss this condition. The upper end of the spring travels downward by 295 mm while the lower end is fixed.

For a method to model the restrained condition of the end coil touching the surface of a ring mount (or any kind of bearing surface) without using any special element like a joining element, please refer to analytical mode 1, ..., 3 in Figs. 5.28, ..., 5.30.

Table 5.13: Element data

| element NO. | node NO. | | section configuration (mm) |
|---|---|---|---|
| 1 | 1 | 2 | round: 6.4 |
| 2 | 2 | 3 | round: 6.4 |
| 3 | 3 | 4 | round: 6.4 |
| 4 | 4 | 5 | round: 6.4 |
| 5 | 5 | 6 | round: 6.4 |
| 6 | 6 | 7 | round: 6.4 |
| 7 | 7 | 8 | round: 6.4 |
| 8 | 8 | 9 | round: 6.4 |
| 9 | 9 | 10 | round: 6.4 |
| ... | ... | ... | ... |
| ... | ... | ... | ... |
| 138 | 138 | 139 | round: 6.4 |
| 139 | 139 | 140 | rectangle: 12.80 × 10.84 |
| 140 | 140 | 141 | rectangle: 12.80 × 9.96 |
| 141 | 141 | 142 | rectangle: 12.80 × 9.40 |
| 142 | 142 | 143 | rectangle: 12.80 × 9.00 |
| 143 | 143 | 144 | rectangle: 12.76 × 8.72 |
| 144 | 144 | 145 | rectangle: 12.68 × 8.56 |
| 145 | 145 | 146 | rectangle: 12.59 × 8.06 |
| 146 | 146 | 147 | rectangle: 12.48 × 6.88 |
| 147 | 147 | 148 | rectangle: 12.36 × 5.36 |

Table 5.14: Data of material characteristics

| | |
|---|---|
| Young's modulus (GPa) | 186.8 |
| shear modulus (Gpa) | 75.3 |

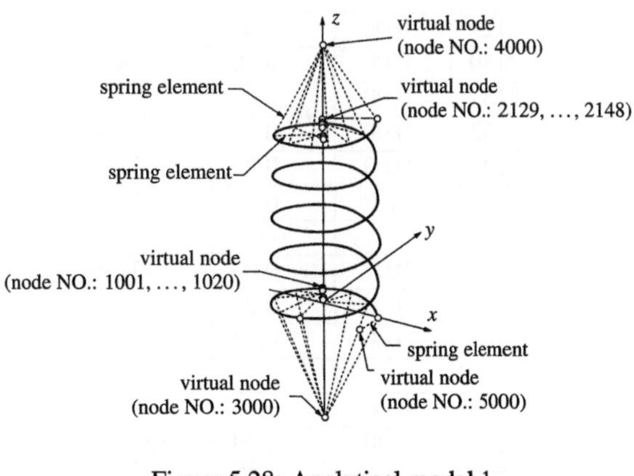


Figure 5.28: Analytical model 1.

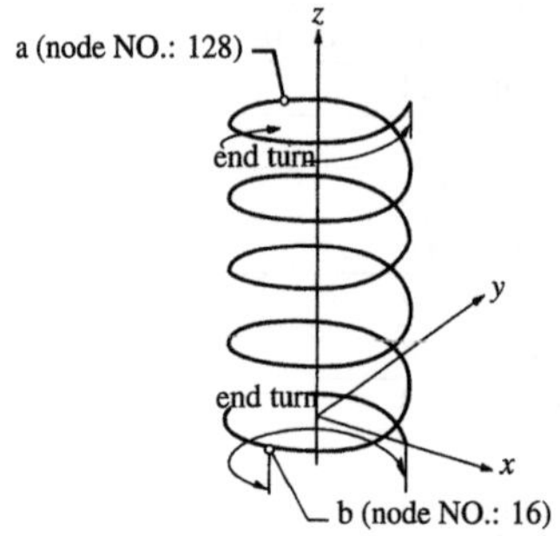


Figure 5.29: Analytical model 2.

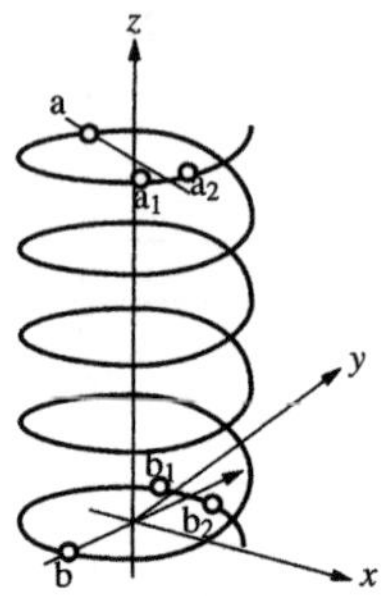


Figure 5.30: Analytical model 3.

Table 5.15: Nodal data of virtual nodes

| node NO. | $x$ (mm) | $y$ (mm) | $z$ (mm) |
|---|---|---|---|
| 1001 | 0.00 | 0.00 | 0.00 |
| 1002 | 0.00 | 0.00 | 0.19 |
| 1003 | 0.00 | 0.00 | 0.45 |
| 1004 | 0.00 | 0.00 | 0.80 |
| 1005 | 0.00 | 0.00 | 1.26 |
| 1006 | 0.00 | 0.00 | 1.85 |
| ... | ... | ... | ... |
| 1018 | 0.00 | 0.00 | 20.20 |
| 1019 | 0.00 | 0.00 | 22.61 |
| 1020 | 0.00 | 0.00 | 25.05 |
| | | | |
| 2129 | 0.00 | 0.00 | 421.09 |
| 2130 | 0.00 | 0.00 | 424.54 |
| 2131 | 0.00 | 0.00 | 427.58 |
| 2132 | 0.00 | 0.00 | 430.08 |
| 2133 | 0.00 | 0.00 | 432.13 |
| 2134 | 0.00 | 0.00 | 433.86 |
| ... | ... | ... | ... |
| 2146 | 0.00 | 0.00 | 435.61 |
| 2147 | 0.00 | 0.00 | 435.69 |
| 2148 | 0.00 | 0.00 | 436.76 |
| | | | |
| 3000 | 0.00 | 0.00 | – 800.00 |
| 4000 | 0.00 | 0.00 | 1236.80 |
| 5000 | 42.92 | –10.00 | 0.00 |

When each analytical result is compared to the experiment, it becomes clear which analytical model is appropriate. This result will be explained later.

The model shown in Fig. 5.28 takes the contact state between surface and end turn as a spring element. In this model, the displacement of both end turns within *x-y* plane is restrained by connecting virtual nodes located on *z*-axis (nodal number 1001, ..., 1020 and 2129, ..., 2148) and nodes on the end turns (nodal number 1, ..., 20 and 129, ..., 148) by spring elements. The rotation of the spring about *z*-axis is restrained by connecting virtual node located on the extension of lower end turn (nodal number 5000) and node at end (nodal number 1) by spring element. The value of *z* coordinate for these virtual nodes is to be equal to the value of the coordinate of corresponding nodes on the coil.

The displacement of both coil ends in *z* direction are restrained by connecting two virtual nodes (nodal number 3000 and 4000), which are located on *z* axis sufficiently far apart from each spring end, and nodes on end turns by spring elements. Broken lines in the figure show the spring elements.

Nodal data (nodal number, coordinate) of the virtual nodes is listed in Table 5.15.

Figures are indicated by *x-y-z* coordinate system shown in Fig .5.25. The spring element data of element number, constituting nodal number and spring constant are shown in Tables 5.16 and 5.17.

The spring constant is to be 5 kN/mm for all the virtual elements in this case. The superposition of virtual element stiffness and the stiffness of helical spring itself constitutes global stiffness. Table 5.18 shows the boundary condition of this model.

Fig. 5.29 shows a model on an assumption that the deformation of end turn is completely restrained by surface. The boundary condition in this case is shown in Table 5.19.

Table 5.16: Element data of the spring element. (for *x-y* restraint)

| element NO. | node NO. | | spring constant (kN/mm) |
|---|---|---|---|
| 1001 | 1 | 1001 | 5.00 |
| 1002 | 2 | 1002 | 5.00 |
| 1003 | 3 | 1003 | 5.00 |
| ... | ... | ... | |
| 1018 | 18 | 1018 | 5.00 |
| 1019 | 19 | 1019 | 5.00 |
| 1020 | 20 | 1020 | 5.00 |
| | | | |
| 2129 | 129 | 2129 | 5.00 |
| 2130 | 130 | 2130 | 5.00 |
| 2131 | 131 | 2131 | 5.00 |
| ... | ... | ... | |
| 2146 | 146 | 2146 | 5.00 |
| 2147 | 147 | 2147 | 5.00 |
| 2148 | 148 | 2148 | 5.00 |

Table 5.17: Element data of the spring element. (for $z$ restraint)

| element NO. | node NO. | | spring constant (kN/mm) |
|---|---|---|---|
| 3001 | 1 | 3000 | 5.00 |
| 3002 | 2 | 3000 | 5.00 |
| 3003 | 3 | 3000 | 5.00 |
| ... | ... | ... | |
| 3014 | 14 | 3000 | 5.00 |
| 3015 | 15 | 3000 | 5.00 |
| 3016 | 16 | 3000 | 5.00 |
| | | | |
| 4129 | 129 | 4000 | 5.00 |
| 4130 | 130 | 4000 | 5.00 |
| 4131 | 131 | 4000 | 5.00 |
| ... | ... | ... | |
| 4146 | 146 | 4000 | 5.00 |
| 4147 | 147 | 4000 | 5.00 |
| 4148 | 148 | 4000 | 5.00 |
| | | | |
| 5000 | 1 | 5000 | 5.00 |

Table 5.18: Boundary condition. (Analytical model 1)

| node NO. | displacement (mm) | | |
|---|---|---|---|
| | $x$ | $y$ | $z$ |
| 1001, ..., 1020 | 0.0 | 0.0 | 0.0 |
| 2129, ..., 2148 | 0.0 | 0.0 | −295.0 |
| 3000 | 0.0 | 0.0 | 0.0 |
| 4000 | 0.0 | 0.0 | −295.0 |
| 5000 | 0.0 | 0.0 | 0.0 |

Table 5.19: Boundary condition. (Analytical model 2)

| node NO. | displacement (mm) | | | rotation (deg.) | | |
|---|---|---|---|---|---|---|
| | $x$ | $y$ | $z$ | $x$ | $y$ | $z$ |
| 16 | 0.0 | 0.0 | 0.0 | 0.0 | 0.0 | 0.0 |
| 128 | 0.0 | 0.0 | −295.0 | 0.0 | 0.0 | 0.0 |

Table 5.20: Boundary condition. (Analytical model 3)

| node NO. | displacement (mm) | | | rotation (deg.) | | |
|---|---|---|---|---|---|---|
| | $x$ | $y$ | $z$ | $x$ | $y$ | $z$ |
| 5 | 0.0 | 0.0 | 0.0 | free | free | free |
| 7 | 0.0 | 0.0 | 0.0 | free | free | free |
| 16 | 0.0 | 0.0 | 0.0 | free | free | 0.0 |
| 128 | 0.0 | 0.0 | −295.0 | free | free | free |
| 137 | 0.0 | 0.0 | −295.0 | free | free | free |
| 139 | 0.0 | 0.0 | −295.0 | free | free | free |

Node No.16 is at the lower end of free turn (point b of Fig. 5.29) and node No.128 is at the upper end of the turn (point a of Fig. 5.29).

Lastly, Fig. 5.30 presents a model on an assumption that end turn has contact with surface at three points. The point a (nodal number 128), $a_1$ (nodal number 137), and $a_2$ (nodal number 139) of the upper end turn and the point b (nodal number 16), $b_1$ (nodal number 7), and $b_2$ (nodal number 5) of the lower end turn are assumed to contact the bearing surface. The points a and b are at the ends of the free turn. The points $a_1$ and $a_2$ are located about 180 degrees opposite from point a around $z$-axis. The points $b_1$ and $b_2$ are the nodes at the similar positions concerning point b. The boundary condition of this analytical model is shown in Table 5.20.

**(2) Postprocessing and analytical accuracy** Nodal displacement, section force and section moment are the result of analysis. Other than that, stress at the element integral point is also calculated in the case that the element stiffness matrix is calculated by numerical integration. However, the effect of coil curvature on stress is not generally considered in beam element prepared for commercially available software packages. To obtain this stress, we must calculate it separately from the section force or the section moment using a technique presented here[15]. To rid ourselves of such inconvenience, a spatially curved beam element having both curvature and torsion has been developed[14].

The technique will start by the transformation of load vector from coordinate system shown in Fig. 5.26 into load vector ($P_\xi, P_\eta, P_s, M_\xi, M_\eta, M_s$) in local coordinate system ($\xi$-$\eta$-$s$ coordinate system) shown in Fig. 5.31.

$\xi$ axis, $\eta$ axis, and $s$ axis are in normal, binormal, and tangential direction, respectively, at an arbitrary point on the centerline of helical wire. Then calculation formula shall be,

$$\sqrt{g}\begin{bmatrix}\tau_\xi \\ \tau_\eta \\ \sigma_s\end{bmatrix} = \begin{bmatrix}\dfrac{P_\xi}{A_{\text{eff}}} \\ \dfrac{P_\eta}{A_{\text{eff}}} \\ \dfrac{P_s}{A}\end{bmatrix} + \begin{bmatrix}-\dfrac{M_s}{I_p}\eta \\ \dfrac{M_s}{I_p}\xi \\ \dfrac{M_\xi}{I}\eta - \dfrac{M_\eta}{I}\xi\end{bmatrix}, \tag{5.9}$$

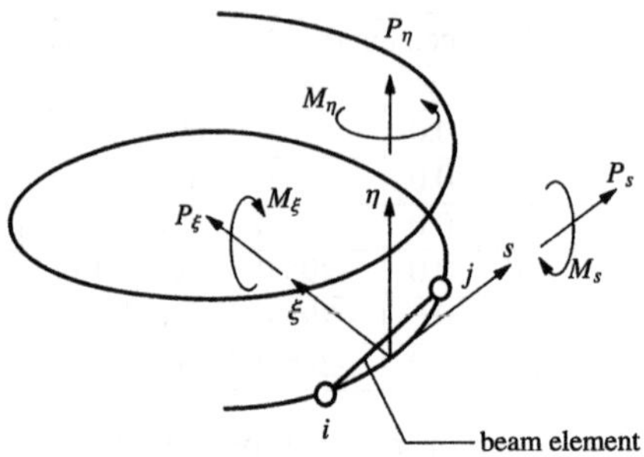


Figure 5.31: Local coordinate.

where $\sqrt{g} = 1 - \kappa\xi$, $\kappa = 1/R(1 + h^2)$, $h = (z_j - z_i)/l$, $z_i$, $z_j$; $z$-coordinate of the element, nodes $i$ and $j$ after deformation, $l$; length of the element, $R$; coil diameter after deformation, $A$; cross-sectional area, $A_{\text{eff}}$; effective section area, $E$; Young's modulus, $I$; moment of inertia of area, $I_p$; polar moment of inertia of area, $G$; shear modulus.

Surface stress expressed by $r$-$\theta$ polar coordinate on the cross-section of wire shown in Fig. 5.32 would be more convenient than that of the one expressed by $\xi$-$\eta$ rectangular coordinates.

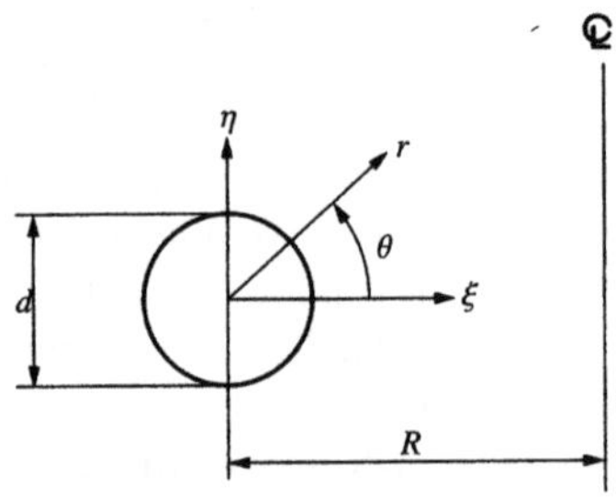


Figure 5.32: Polar coordinate expression of cross section

In this case, it is necessary to replace $\xi$ and $\eta$ in (5.9) with $d{\cdot}\cos\theta/2$ and $d{\cdot}\sin\theta/2$, respectively.

Fig. 5.33 and Fig. 5.34 show an analytical result for load-deflection characteristic and tangential spring constant.

Fig. 5.34 shows that the tangential spring constant of model 2, i.e., the model in which the deformation of end turn is assumed to be completely restrained by bearing surface, shows the highest value. The spring constant obtained by model 2 is greater than the experimental value by 38% or more in the range of small deflection and by approximately 12% even at a deflection of 150 mm. It is apparent from this fact that the influence of end turn deformation cannot be neglected. Alternately, there is

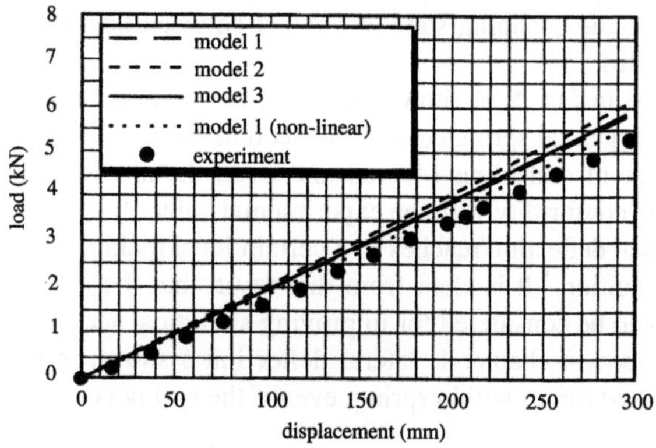


Figure 5.33: Load-deflection characteristic.

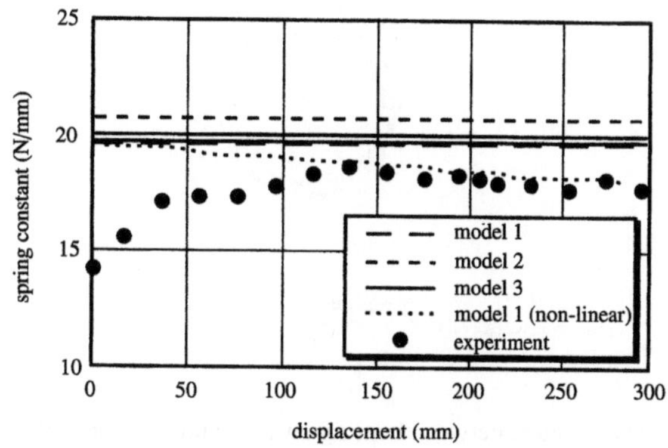


Figure 5.34: Tangential spring constant.

almost no difference to the analytical result of the tangential spring constant between model 1 and 3. At the deflection of 150 mm, the difference between analytical result and experimental value is roughly 6%.

The reason why an analytical spring constant is higher than experimental one can be attributed to the insufficient modeling of contact state between end turn and bearing surface. For springs with closed taper ends, in particular, the part of end coil (shadowed area) in Fig. 5.35 is not fully pressed against bearing surface.

Refer to table 5.18 and 5.20 for the boundary conditions set for models 1 and 3.

As can be seen in Fig. 5.34, experimental tangential spring constant has a tendency to increase with deflection in the range of small deflection. There is a probable mechanisms to explain this. First, the tip of pigtail end starts touching bearing surface and the number of contact coils increases as the spring is compressed. Secondly, the end turn of closed taper starts touching next turn with the increase of applied load. To improve the analytical accuracy, it is necessary to make a model closely simulating the changing state of contact between bearing surface and end turn or mutual wires.

The dotted line in Fig. 5.34 shows the result of geometrical nonlinear analysis of model 1, the tangential spring constant is decreasing with the increase of deflection. In the range of deflection 140 mm and over, the analytical value begins to demonstrate solid conformity with the experimental value. It seems that contact status get stable when deflection reaches to a certain level making the situation closer to theoretical condition. It is clear from this result that the effect of geometrical nonlinearity should be considered for improving analytical accuracy.

As was described in chapter 1, lateral deflection $\delta$ shown in Fig. 5.36 will be caused by combined forces within spring, even if the spring is compressed squarely to end surface.

The spring designed to bear high stress to save weight has a tendency to show a large lateral deflection owing to its lesser number of coils or large pitch angle. Room

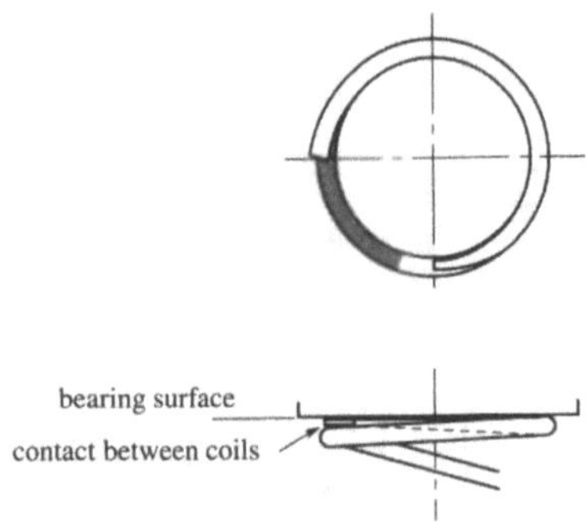


Figure 5.35: State of contact between bearing surface and end coil. (closed taper end)

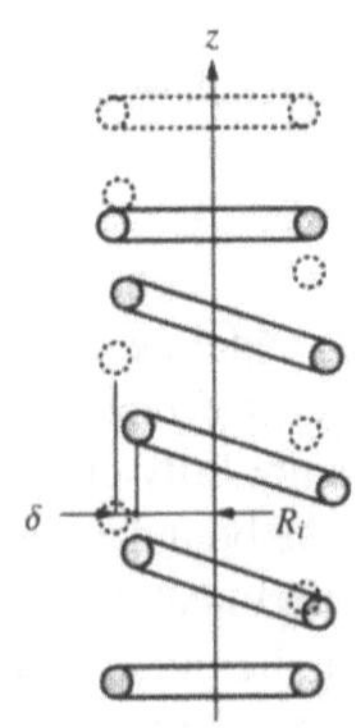


Figure 5.36: Lateral deflection of a helical spring.

for springs in machines is narrowing in the quest for small sized instruments. From these circumstances, it is important to eliminate the possibility of interference between springs and peripheral parts. Consequently, the need for an advance analysis of the lateral deflection[18] is inevitable.

As was described in chapter 1, conventional design formulas do not include the restraining condition (lateral force, moment) on end turn. In case of FEM, to the contrary, the multiple degrees of freedom is reserved automatically through the modeling of spring by beam element and the lateral deflection of spring is to be carried out easily. Let us grasp the analytical accuracy of lateral deflection based on the result of linear analysis for the three models studied here.

Fig. 5.37 shows coil radius variation when the spring is axially deflected by 295 mm.

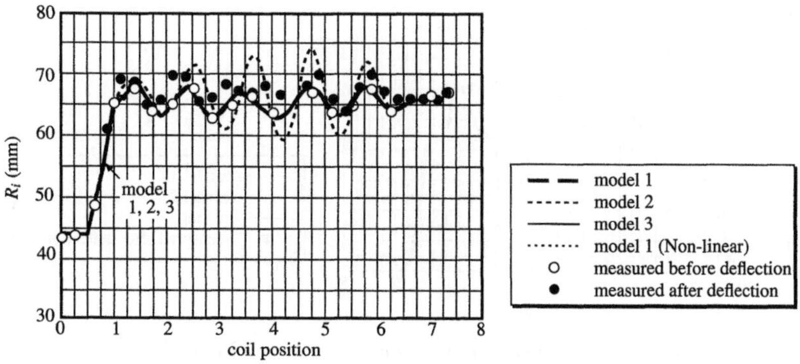


Figure 5.37: Radius of coil after deformation.

The abscissa of the figure shows the position of coil measured from the tip of lower end turn (pig tail end side) and the ordinate shows the coil radius $R_i$ shown in Fig. 5.36. Bold broken line, dotted line, and solid line in Fig. 5.37 show the result of model 1, model 2, and model 3, respectively, and dark circles show the experimental values. Open circles show the coil radius in unloaded condition. The reason for fluctuation in the coil radii in unloaded condition is the initial occurrence of lateral deflection by presetting. As can be understood from the figure, the magnitude of fluctuation cannot be ignored in interference check. The change of shape caused by presetting is incorporated into the analytical model because the measured value was used for the shape data in the present analysis.

When we apply FEM analysis for checking space interference, however, the estimation of shape change caused by presetting, i.e., elasto-plastic analysis, is also necessary. The analysis of presetting for helical spring will be studied in chapter 6.

As is evident from the figure, the experimental value shows that coil radius increases when spring is compressed, but the analytical result for each model shows that coil radius after deformation is almost identical to pre-deformation (the analytical results of models 1 and 2 almost overlap onto the results of model 3.). In the geometrical nonlinear analysis of model 1, on the other hand, the change of coil ra-

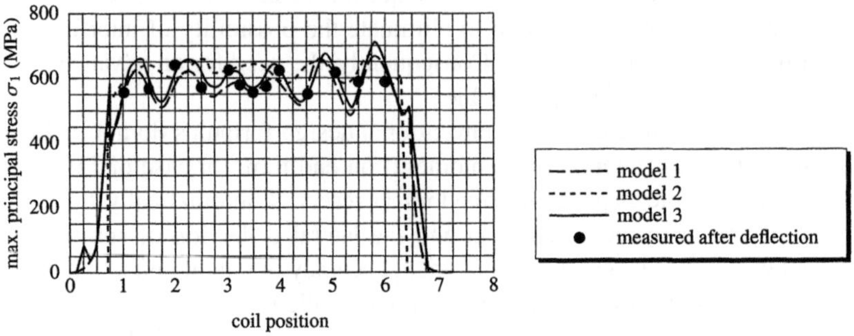


Figure 5.38: Stress distribution on the outside of coil (Load: 3.64 kN).

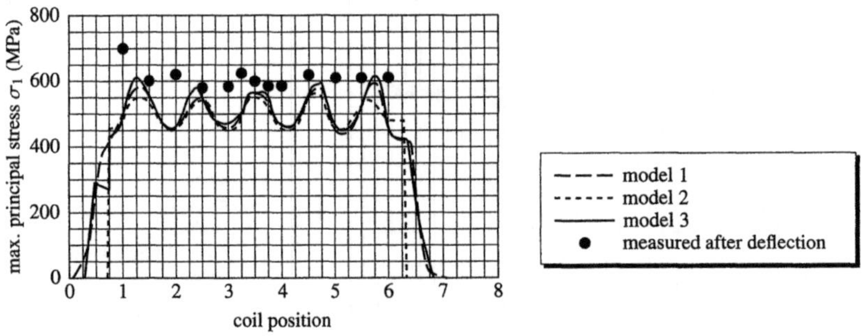


Figure 5.39: Stress distribution on the inside of coil Load: 3.64 kN).

dius is evident. The change of coil radius is considerably larger than the experiment. The reason for the above discrepancy may be that the change of the contact state between bearing surface and end turn or inter coil contact during the compression is not mathematically stated with enough precision in this model. The proper model of the contact state with geometrical nonlinear analysis is essential for the accurate estimation of lateral deflection.

Lastly, let's examine stress. Fig. 5.38 and Fig. 5.39 show the distribution of principal stress on the outside of coil (on the surface at $\theta = 180°$ in Fig. 5.32) and on the inside of coil (on the surface at $\theta = 0°$ in Fig. 5.32), respectively, under an applied load 3.64 kN.

The abscissa of the figure shows the position of coil counted from the tip of lower end turn (pig tail end side) and the ordinate shows the maximum principal stress. The maximum principal stress was calculated from $\tau_\xi$, $\tau_\eta$, and $\sigma_s$ obtained by (5.9). Bold broken line, dotted line, and solid line in the figure correspond to model 1, model 2, and model 3, respectively. The dark circles show experimental value. Fig. 5.40 and Fig. 5.41 show similar results under an applied load of 5.26 kN.

Fig. 5.38 shows that the experimental values of model 1 and 3 have similar data for stress distribution on the outside diameter of coil. For model 2, the coil position with the maximum principal stress shows some variance from the experimental

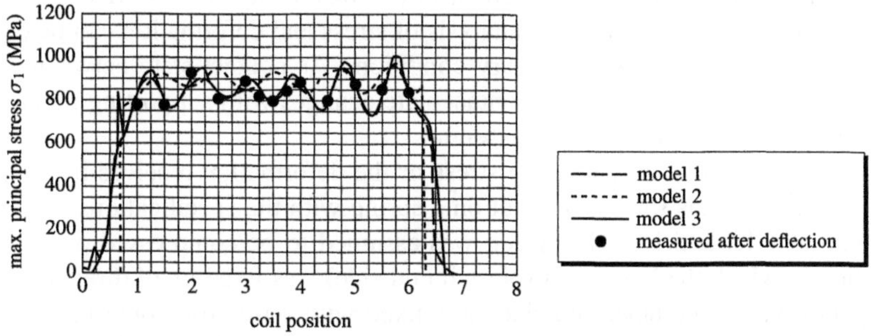


Figure 5.40: Stress distribution on the outside of coil (Load: 5.26 kN).

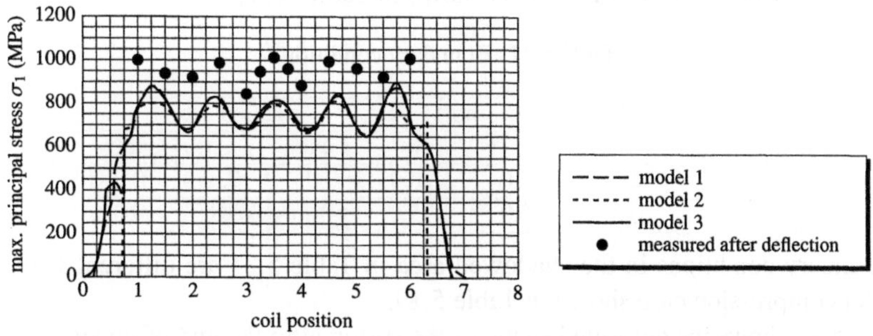


Figure 5.41: Stress distribution on the inside of coil (Load: 5.26 kN).

value, although the maximum value of the maximum principal stress is almost the same as model 1 and 3. This is apparently caused by the difference in restraining condition of end turn. As can be seen in Fig. 5.40, the stress distribution on the outside of coil shows similar results even when the applied load is increased. For the stress distribution on the inside of coil, the calculated stress fluctuation shows a larger amplitude than the one in experiment, as shown in Fig. 5.39. The maximum stress, however, shows no large difference from the experimental value, with an exception of the neighborhood of the first coil. However, as the applied load is increased, maximum principal stress and its fluctuation of the experimental value becomes larger than those of the calculation as is shown in Fig. 5.41. The reason for this phenomenon is assumed that the change in coil diameter during compression is not taken into account, and eccentric oblique load is not evaluated accurately in the linear analysis.

**(3) Effect of supporting method** The helical spring is not always used in the state of parallel compression. The usage could often be in compressed setting with offset, or a knee action setting with rotational displacement. Sometimes the helical spring is used with a support allowing free rotation. If the supporting method is different, stress and deformation will also be different.

The finite element method is excellent in its flexible capability to handle such a variety of supporting methods. An analysis of stress and deformation can be easily carried out just by changing boundary condition. As contact state between bearing surface and coil, or inter-wire contact state changes with supporting method, a proper modeling of contact state is essential.

Let's examine what influence the stress or the deformation receives by support method. The analysis was carried out on a spring with the specifications shown in Fig. 5.23 and Table 5.11. The spring was assumed to be compressed on support condition in which the lower seat (pig tail end side) was fixed in all directions and the upper seat (closed taper end side) was fixed with a free rotational jig. As the precise modeling of contact state between bearing surface and end coil on the above supporting condition is not so easy, the model of Fig. 5.29 was used for an expedient. The boundary condition of this case is shown in Table 5.21.

Table 5.21: Boundary condition

| node NO. | displacement (mm) | | | rotation (deg.) | | |
|---|---|---|---|---|---|---|
| | $x$ | $y$ | $z$ | $x$-axis | $y$-axis | $z$-axis |
| 16 | 0.0 | 0.0 | 0.0 | 0.0 | 0.0 | 0.0 |
| 128 | 0.0 | 0.0 | −295.0 | free | free | 0.0 |

Boundary conditions in the shadowed area in Table 5.21 are different from the parallel compression case shown in Table 5.19.

Fig. 5.42 shows the deformed shape of the spring after the end of upper free turn (nodal number 128) was displaced downward by 69 mm.

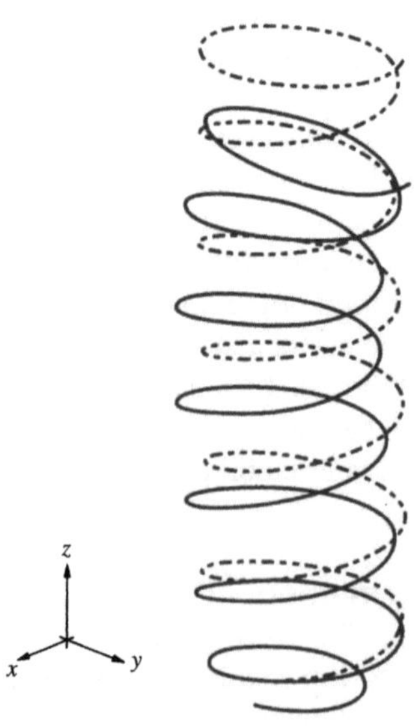


Figure 5.42: Deformed shape of spring.

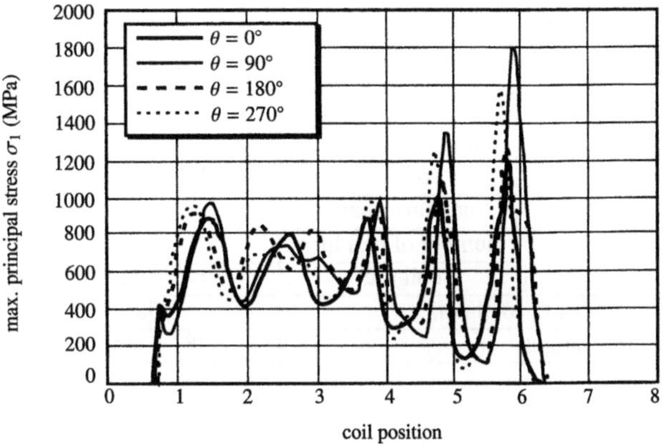


Figure 5.43: Stress distribution.

The upper part of the spring looks in a slanted position, as the upper seat was set free to rotate about its $x$ and $y$ axis. Fig. 5.43 shows the distribution of maximum principal stress when the end of upper free turn is displaced downward by 253.5 mm.

The abscissa shows the position of coil measured from the tip of lower end (pig tail end side) and the ordinate shows maximum principal stress. The notation $\theta$ in the figure is an angle in the cylindrical coordinate system shown in Fig. 5.32. The load necessary to compress by 253.5 mm is 5.26 kN. When the comparison is made against Fig. 5.40 and Fig. 5.41, the result shows that stress in the neighborhood of upper end turn is very high when upper seat can rotate freely about the $x$ and $y$ axis.

As the contact state between bearing surface and end coil is not defined properly enough, the analytical accuracy of stress and deformed shape are not well established; Nevertheless, the effect of support condition on the stress or the deformation could be seen easily by that analysis.

**(4) Effect of coil contour** The contour of the helical compression spring is widely variable: conical, barrel and hourglass-shaped. Corresponding to each of these shapes, various calculating formulas have been devised and utilized. The finite element method, on the other hand, can be applicable to springs of any shape, simply by revising data on shape, restraining condition, and load condition following predetermined procedure and method.

An example of this analysis for a conical spring will be presented here. A conical spring is a spring that has nonlinear spring characteristics through the positive use of contact between wires. To analyze stress or spring characteristics accurately, appropriate modeling of contact condition between wires is a key issue. However, this demo is carried out assuming no contact in order to take a casual glance at the effect of coil shape on stress.

Table 5.22 shows the specification of the conical spring used in the analysis.

Table 5.22: Specification of the analytical conical spring

| | | |
|---|---|---|
| wire diameter | | $\phi$ 11.5 mm |
| material | | SUP7 (JIS G 4801) |
| large dia. | end turn shape | open |
| | number of end turn | 0.75 |
| | coil dia. | 156.2 mm |
| small dia. | end turn shape | open |
| | number of end turn | 0.75 |
| | coil dia. | 122.6 mm |
| number of total coil | | 5.50 |
| free height | | 381.5 mm |

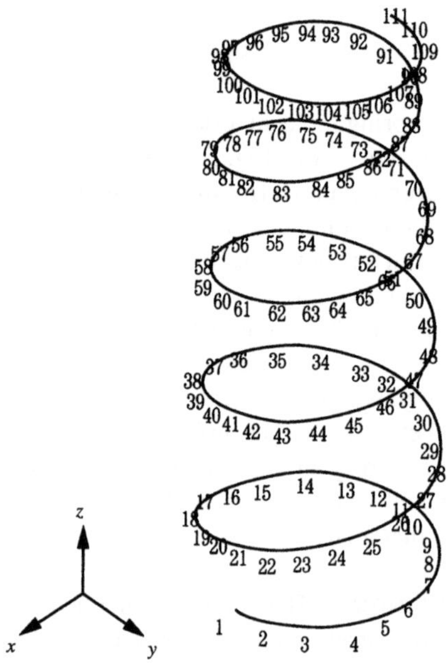


Figure 5.44: An analytical model of conical spring.

The analytical model is shown in Fig. 5.44. We will skip over the input data.

The numerals in the figure are nodal numbers. Supporting condition is parallel compression and the boundary condition follows model 3 in Fig. 5.30.

Fig. 5.45 shows stress distribution when compressed by 248.6 mm keeping end parallel condition.

The abscissa shows the position of the coil measured from the tip of lower end and the ordinate shows maximum principal stress. The notation $\theta$ in the figure is an angle in the cylindrical coordinate shown in Fig. 5.32. The stress shows a wavy

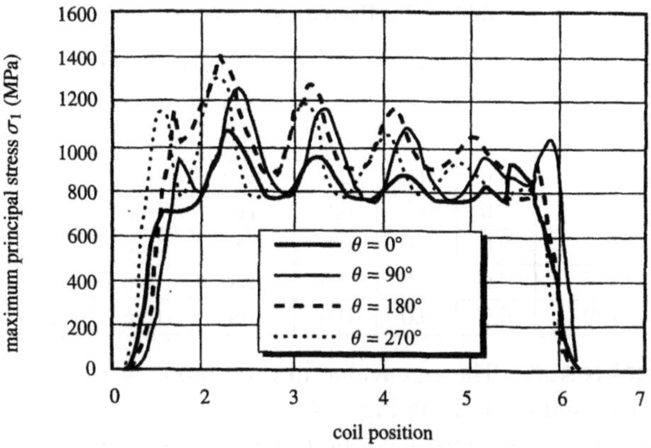


Figure 5.45: Stress distribution in analytical conical spring.

distribution in a way similar to cylindrical spring.

### 5.3.2 Analysis of Eigenvalue

The stress in a helical spring induced by a dynamic load such as impact or vibration is higher than those caused by a static load. When the applied frequency coincides with the natural frequency of the spring, the phenomenon of resonance, so-called surging, takes place. It often becomes the cause of fracture. In the design of springs like engine valve springs that receive a dynamic load, a careful examination of natural frequency or dynamic stress is vital.

When a coil spring is compressed, it deforms not only in longitudinal direction but also in lateral direction. The designing formula[19] which is generally used for the calculation of natural frequency or dynamic stress is applicable only to longitudinal vibration and contact between wires is not considered at all. As a result, we have no option but to depend on experimental means when we want to know accurate dynamic stress or natural frequency.

The analysis of eigenvalue or dynamic stress that considers contact between wires will be discussed in chapter 6. An analytical example of eigenvalue excluding contact between wires will be presented here briefly.

**(1) Analytical model** The spring shown in Fig. 5.23, Table 5.11, and the analytical model of Fig. 5.30 will be discussed here. The eigenvalue and the eigenvector of the helical spring can be easily analyzed by obtaining additional information about the mass density of material. In this case, the mass density was assumed to be $7.8\,\text{kg/mm}^3$. The boundary condition is shown in Table 5.23.

The eigenvalue at zero load will be given in following clause.

Table 5.23: Boundary condition

| node NO. | displacement (mm) | | | rotation (deg.) | | |
|---|---|---|---|---|---|---|
| | $x$ | $y$ | $z$ | $x$-axis | $y$-axis | $z$-axis |
| 5 | 0.0 | 0.0 | 0.0 | free | free | free |
| 7 | 0.0 | 0.0 | 0.0 | free | free | free |
| 16 | 0.0 | 0.0 | 0.0 | free | free | 0.0 |
| 128 | 0.0 | 0.0 | 0.0 | free | free | free |
| 137 | 0.0 | 0.0 | 0.0 | free | free | free |
| 139 | 0.0 | 0.0 | 0.0 | free | free | free |

**(2) Analytical result** The eigenvalue in each mode were calculated up to the eighth order. The result is shown in Fig. 5.46.

As can be seen from the figure, the vibrational modes of the helical spring corresponds not only to longitudinal vibration but also to torsional vibration about central axis, bending vibration, as well as coupled vibration.

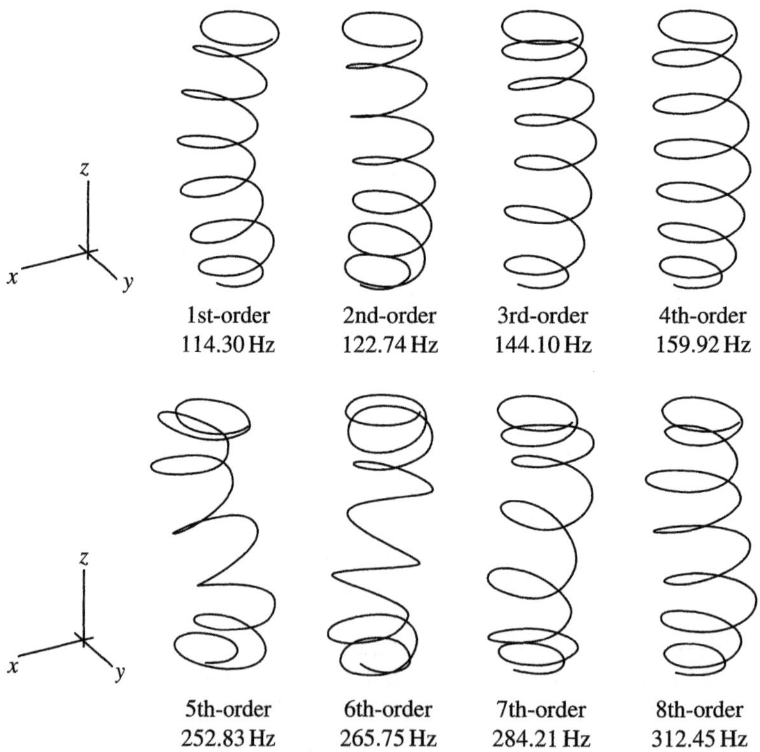


Figure 5.46: Vibrational modes.

# 5.4 HELICAL EXTENSION SPRINGS

Since a helical extension spring is wound to have a negative pitch, it has an initial tension force. Therefore, the load-deflection characteristic of the spring takes a pattern, like the one shown in Fig. 5.47 and the spring does not deflect until the load is larger than the initial tension load.

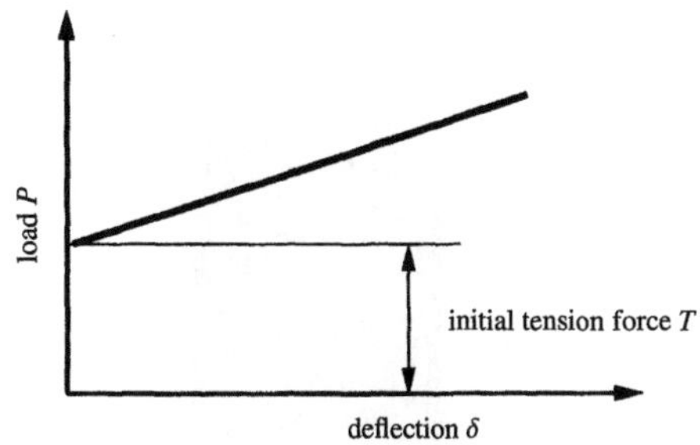


Figure 5.47: Load-deflection characteristic of a helical extension spring.

It is necessary to define a calculation model with a negative pitch when analyzing the load-deflection characteristic of a helical extension spring by FEM. Although the close analysis of such a nonlinear characteristic should be carried out by following a principle described in Paragraph 3.3.2, we will skip initial tension for the benefit of a quick grasp of this spring.

The method of modeling is essentially the same as that of the helical compression spring, since the negative pitch is not taken into account. At the ends of the extension spring, hooks are formed to transmit tensile load to the coil. Therefore, it is necessary to model the hook with finite elements. The analysis is easier than in the case of the helical compression spring since factors of nonlinearity in the boundary condition can be excluded in the analysis.

An analytical example is briefly presented here.

**(1) Outline of an analytical model** The shape, the dimensions, and the specification of the analytical helical spring are shown in Fig. 5.48 and Table 5.24.

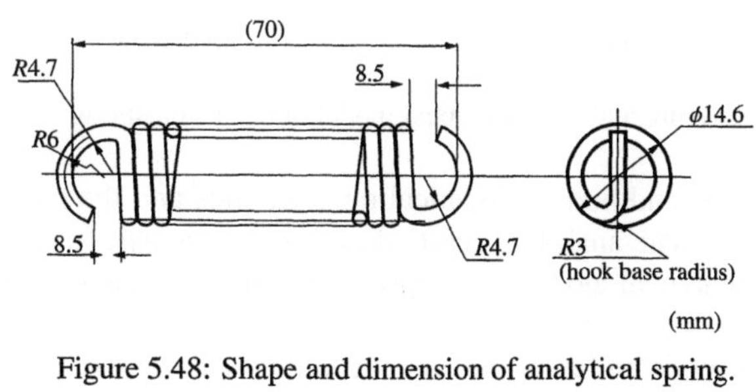


Figure 5.48: Shape and dimension of analytical spring.

Table 5.24: Specification

| | |
|---|---|
| wire dia. (mm) | $\phi$ 2.6 |
| mean coil dia. (mm) | $\phi$12.0 |
| total number of coil | 15.5 |

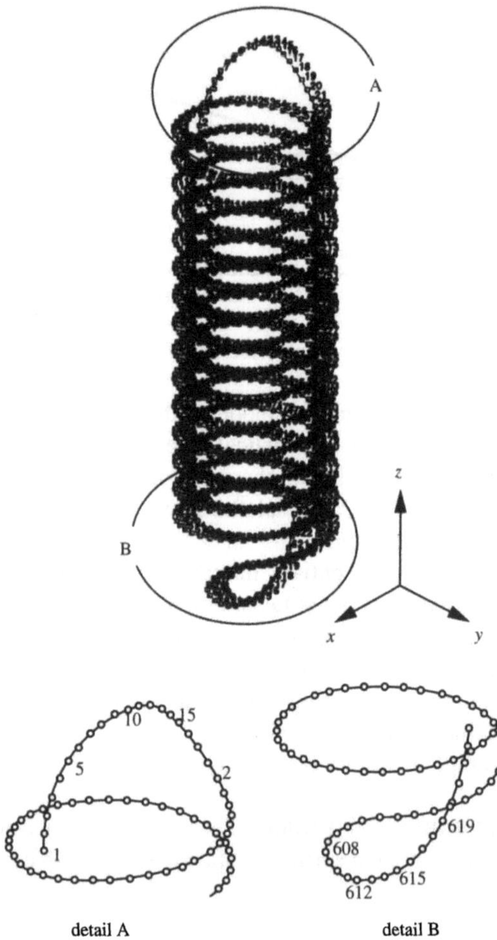


Figure 5.49: An analytical model of extension spring.

The shape of the end hook is identical on both ends and the radius at the base of hook is 3.0 mm. The analysis is carried out under a tensile load of 400 N. Fig. 5.49 shows the analytical model of the spring. The numerals in the figure are nodal numbers.

The spring is modeled, as in the case of the helical compression spring, with

Table 5.25: Boundary condition

| node NO. | displacement | | | rotation | | |
|---|---|---|---|---|---|---|
| | $x$ | $y$ | $y$ | $x$-axis | $y$-axis | $y$-axis |
| 10 | Fixed | Fixed | Free | Free | Free | Free |
| 615 | Fixed | Fixed | Fixed | Fixed | Free | Fixed |

Table 5.26: Load condition

| node NO. | $P_x$ (N) | $P_y$ (N) | $P_z$ (N) |
|---|---|---|---|
| 10 | 0.0 | 0.0 | 400 |

three-dimensional two-nodal beam elements. Young's modulus is assumed to be 205.8 GPa and Poisson's ratio 0.315. The boundary condition and the load condition are shown in Table 5.25 and Table 5.26, respectively.

Node number 10 and 615 are shown in detail A and B of Fig. 5.49.

**(2) Analytical result** Fig. 5.50 shows the deformed shape of the spring.

The deflection of node 10 in the $z$ direction is 23.96 mm and nearly equal to the

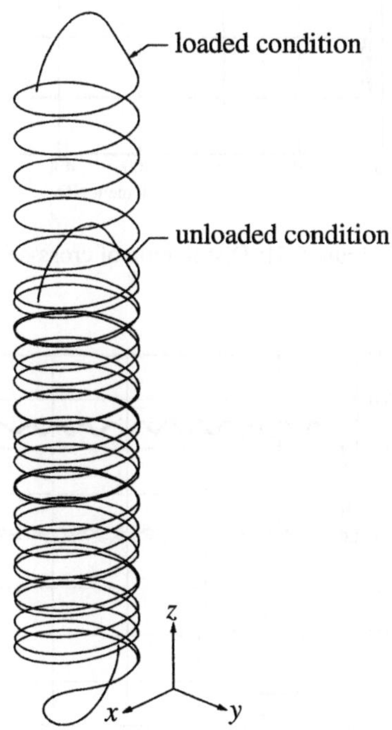


Figure 5.50: Deformed shape.

result ($\delta = 23.92$ mm) using the conventional design formula:

$$\delta = \frac{64PR^3n}{Gd^4}, \tag{5.10}$$

where $R$ and $d$ are the mean coil diameter and the wire diameter respectively. $n$ is the number of coils between hooks. $G$ denotes the shear modulus and $P$ the applied load.

Fig. 5.51 and Fig. 5.52 show cross-sectional force and cross-sectional moment, respectively.

It should be noted that the direction of the second axis in the element coordinate system is orthogonal to $z$-axis in the global coordinate system and to the basic vector $t$ ($s$ axis) in element coordinate system. The $s$ axis in the element coordinate

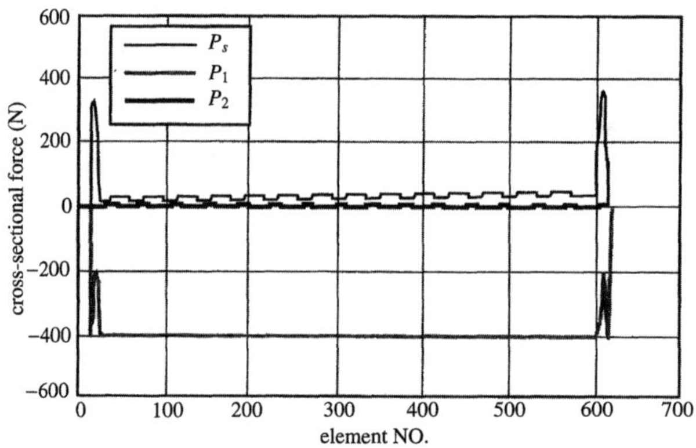


Figure 5.51: Distribution of cross-sectional force.

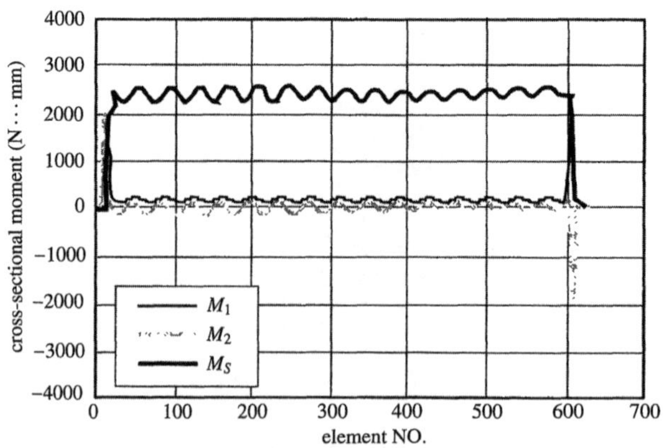


Figure 5.52: Distribution of cross-sectional moment.

system coincides with the axial direction of beam element and its positive direction is the direction of the node number increment. The stress in the coils and the bent part of the hook is calculable by (5.9) transforming cross-sectional force and cross-sectional moment into load vectors $(P_\xi, P_\eta, P_s, M_\xi, M_\eta, M_s)$ in the local coordinate system shown in Fig. 5.31. This calculation requires specialized postprocessing software. We will skip this process in this section.

Large stress concentration tends to be generated at the base curvature of hook, and the cause of fracture is often attributed to this stress concentration. Hence, the examination of stress at the point of the hook is crucial to the design of an extension spring. At the hook section, however, a very complicated stress due to bending and torsion makes an accurate calculation so complex that the formula has not been clearly defined. A reference paper[20] citing Wittrick has achieved a fair degree of accurate analysis on hook stress, however, it is difficult for an average engineer to manipulate the formula. This is the reason why the stress calculation remains at the level of an approximate calculation using Wahl's formula[21]. In the not too distant future, FEM will undoubtedly be seen as an optimum analytical method for solving such problems[22]. Cross-sectional force or cross-sectional moment can be obtained with no effort by FEM software on the market, however, the calculation of stress concentration at the hook bent portion is no easy task, even for sophisticated software. For the present, this may be the limit of general-purpose FEM software.

## 5.5 HELICAL TORSION SPRINGS

A helical torsion spring carries a torsional moment about its coil center axis. Owing to bending moment, a numerical formula is different from those for the helical extension or compression spring to calculate spring characteristic or stress. An advantage of FEM analysis is that, spring characteristic or stress can be calculated by the same procedure as a helical compression or extension spring. As was mentioned in section 5.3, the spatially curved beam element that incorporates curvature and torsion factor is not involved in the commercial FEM software. Stress shall be calculated separately from cross-sectional force and the cross-sectional moment taking curvature and torsion into account.

As discussed in Paragraph 3.3.2, a helical torsion spring could be wound into a solid, similar to the helical extension spring, and hysteresis is generated in spring characteristic. To analyze a solid wound spring with close accuracy, it is necessary to construct a model that can express contact or friction between coils. When the number of coil is large, an analysis of buckling is required in some cases. As a introductory presentation, a simple description based on an analytical example will be presented here, ignoring a contact and buckling evaluation.

**(1) Analytical model** Shape, dimensions, and specification of an analytical helical torsion spring are shown in Fig. 5.53 and Table 5.27.

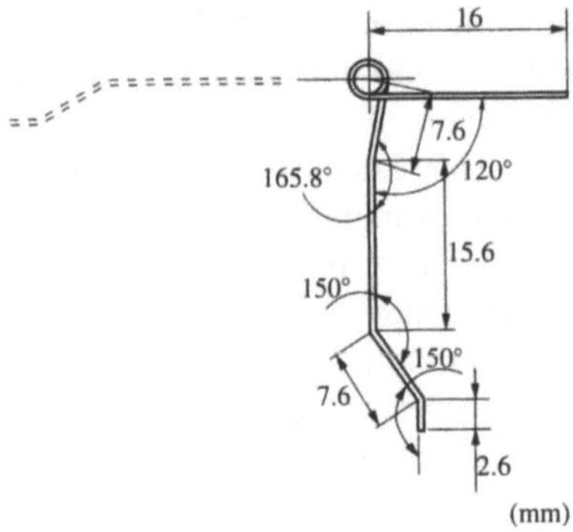


Figure 5.53: Shape and dimension of an analytical helical torsion spring.

Table 5.27: Specification

| | |
|---|---|
| wire dia. (mm) | $\phi$ 0.5 |
| mean coil dia. (mm) | $\phi$ 3.5 |
| number of effective coil | 5.8 |

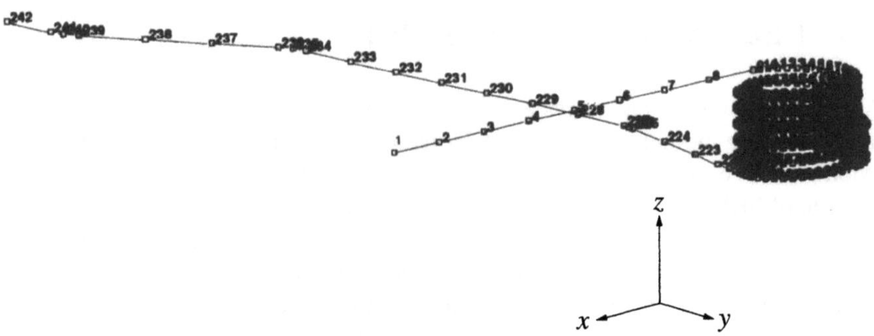


Figure 5.54: An analytical model.

Fig. 5.54 shows the analytical model of the spring. Numerals are node numbers. The spring is modeled with three-dimensional two-nodal beam elements. The direction of second axis in element coordinate system is orthogonal to $z$-axis in the global coordinate system and to basic vector $t$ ($s$ axis) in element coordinate system. The $s$ axis in element coordinate system coincides with the axial direction of beam element and its positive direction is the direction of node number increment. Young's modulus equals 205.8 GPa and Poisson's ratio 0.315.

Displacement boundary condition is shown in Table 5.28. In this case, forced displacement condition is equivalent to about 28.4 degrees of angular displacement.

Table 5.28: Displacement boundary condition

| node NO. | displacement (mm) | | | rotation (rad.) | | |
|---|---|---|---|---|---|---|
| | $x$ | $y$ | $z$ | $x$-axis | $x$-axis | $x$-axis |
| 1 | 0.0 | 0.0 | 0.0 | 0.0 | 0.0 | 0.0 |
| 2 | 0.0 | 0.0 | 0.0 | 0.0 | 0.0 | 0.0 |
| 3 | 0.0 | 0.0 | 0.0 | 0.0 | 0.0 | 0.0 |
| 4 | 0.0 | 0.0 | 0.0 | 0.0 | 0.0 | 0.0 |
| 5 | 0.0 | 0.0 | 0.0 | 0.0 | 0.0 | 0.0 |
| 6 | 0.0 | 0.0 | 0.0 | 0.0 | 0.0 | 0.0 |
| 7 | 0.0 | 0.0 | 0.0 | 0.0 | 0.0 | 0.0 |
| 8 | 0.0 | 0.0 | 0.0 | 0.0 | 0.0 | 0.0 |
| 242 | −15.0 | free | 0.0 | free | free | free |

**(2) Analytical result** Fig. 5.55 shows the deformed shape.

Fig. 5.56 and Fig. 5.57 show cross-sectional force and cross-sectional moment in the beam element. In the figures, abscissa is the number of element.

The bending moment $M_1$ in Fig. 5.57 is nearly the bending moment about $z$-axis. It is obtained by the conventional design formula:

$$M = \frac{Ed^4}{64Dn}\phi, \tag{5.11}$$

where, $E$ is Young's modulus, $d$ is wire diameter, $D$ is mean coil diameter, $n$ is the effective number of coils, and $\phi$ is angular displacement. The bending moment is calculated as 4.9 N·mm based on this formula ($\phi$ = 28.4 degree). Although the bending moment $M$ is assumed to be uniform in (5.11), FEM analysis shows periodical

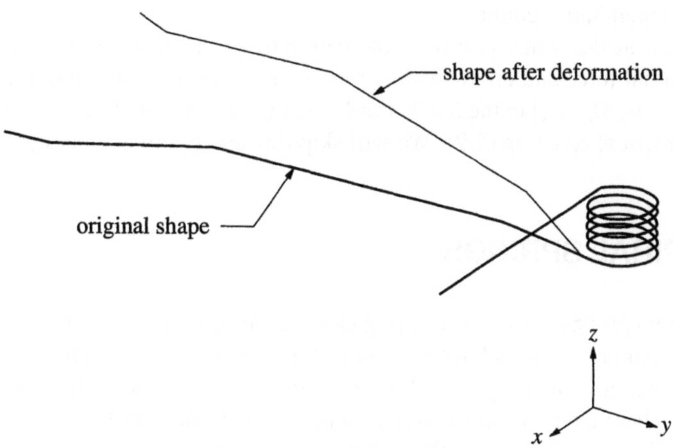


Figure 5.55: Deformed Shape

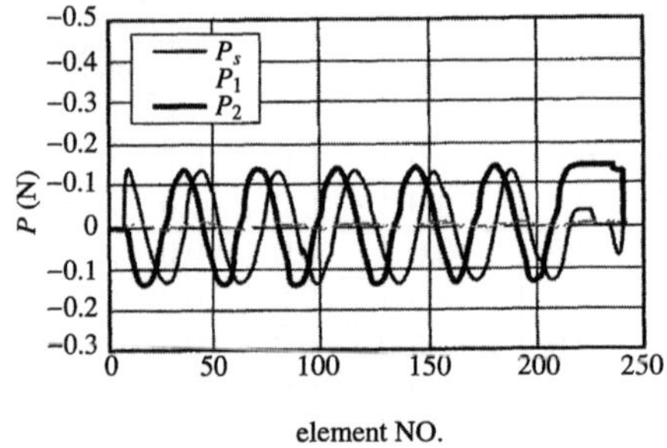


Figure 5.56: Distribution of cross-sectional force.

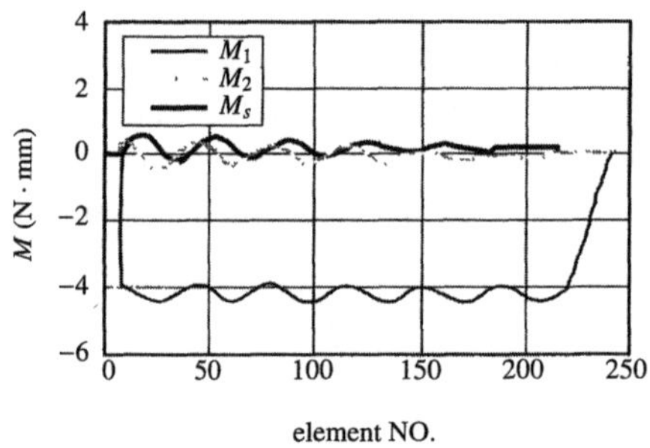


Figure 5.57: Distribution of cross-sectional moment.

fluctuation within a range of 3.93–4.42 N·mm. The bending moment of conventional formula gives larger result than FEM analysis, because the deformation of the arm is not taken into account.

The stress at the coiled part or at the bent part of the arm can be calculated by cross-sectional force and cross-sectional moment being converted into load vectors ($P_\xi, P_\eta, P_s, M_\xi, M_\eta, M_s$) in the local coordinate system shown in Fig. 5.31 by applying the numerical equation (5.9). We will skip this process in this paragraph.

## 5.6 SPIRAL SPRINGS

There are two groups in the spiral spring category according to whether the adjacent coils make contact with each other or not. A spiral spring in the shape of letter S, which belongs to contact type, is housed within a barrel after a twofold winding operation. There are two spirals wound in reverse direction for this type. To obtain the relation between torque and the number of turns or to discern stress distribution on the surface of steel strip of contact type spiral spring, it is necessary, as was described in Paragraph 3.4.2, to take friction among coils into account and to define the shape of spring at a free condition after forming. The analysis of initial stress when the spring is housed in the barrel is also necessary[23, 24]. In other words, material nonlinearity must be taken into account in the analysis, in addition to contact, friction, and geometrical nonlinearity. In the case of no-contact type spiral spring, only geometrical nonlinearity shall be considered. A non-contact spiral spring will be presented here.

**(1) Outline of an analytical model** The shape and the dimension of the spiral spring to be analyzed are shown in Fig. 5.58.

The thickness and the width of strip are 2.5 mm and 5.0 mm, respectively.

As is shown in the figure, the spring is fixed on a center shaft, and is loaded by turning the outer side with a pin. The cross-sectional shape of the coil is not a perfectly rectangular but has some roundness on its side-edge, as is shown in Fig. 5.59.

The four-nodal shell element will be used here for model construction.

Fig. 5.60 and Fig. 5.61 show the analytical models based on four-nodal shell elements.

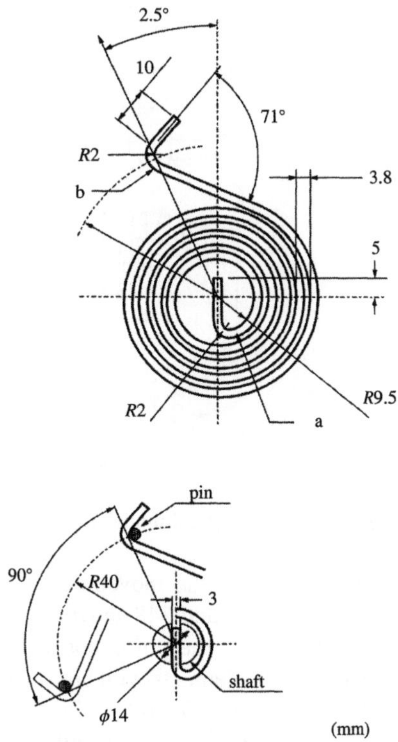


Figure 5.58: Shape and dimension of analytical spiral spring.

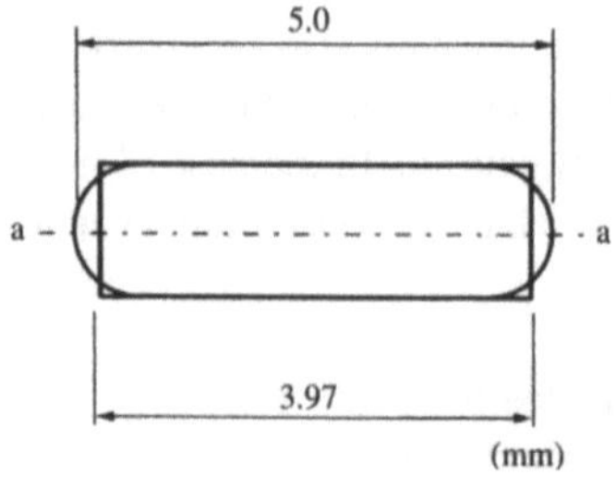


Figure 5.59: Cross-sectional shape of material.

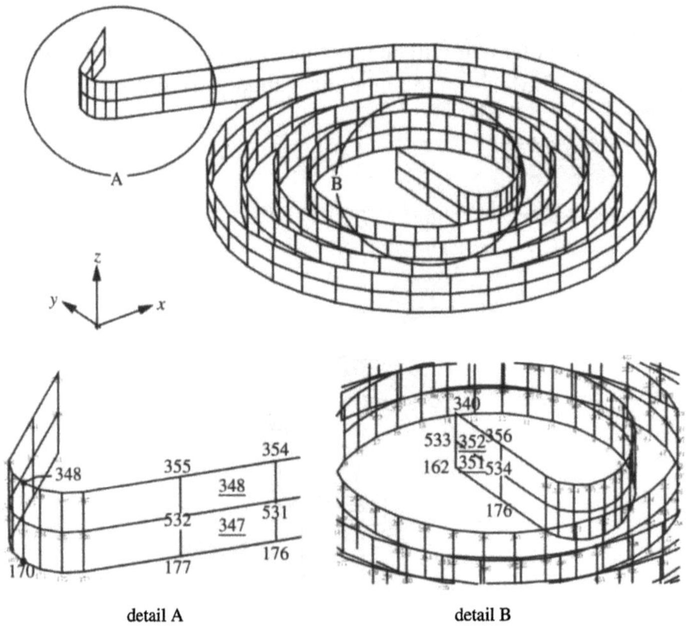


Figure 5.60: Analytical model 1.

In the first analytical model, Fig. 5.60, the movement of pin is expressed by giving a forced displacement to nodes 170 and 348 shown in detail A of Fig. 5.60. In analytical model 2 shown in Fig. 5.61, the pin is connected to a vertual truss elements which connect to vertual beam elements with considerably large stiffness. Nodes 1001, 1002, 2001, and 2002 shown in detail B of Fig. 5.61 are virtual nodes.

The numerals indicated in both figures are nodal number and element number (underlined). In this model, the nodal coordinate was decided so that the cross-section of material would be equivalent to a rectangular shape, as is shown in

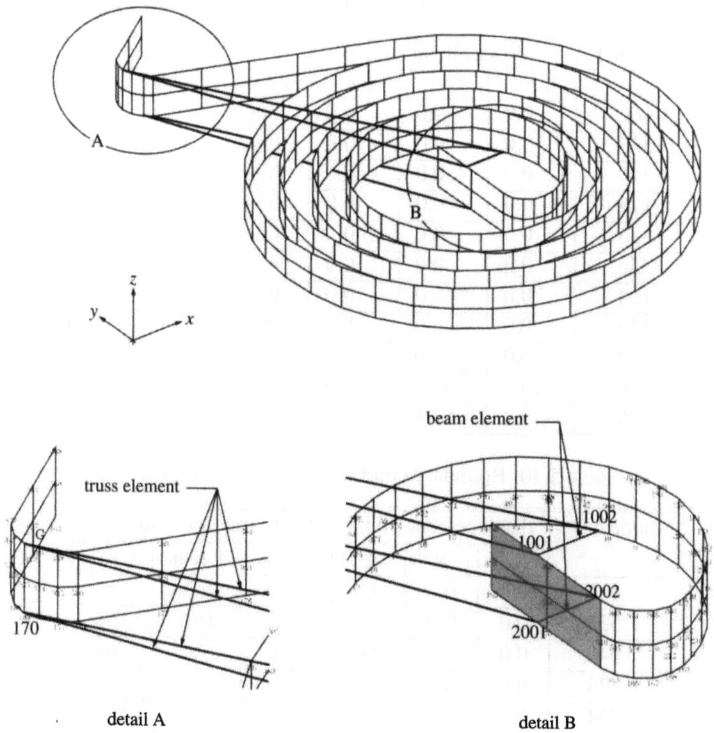


Figure 5.61: Analytical model 2.

Fig. 5.59. The moment of inertia of area about a-a axis is equivalent to that of the actual shape.

To define the shell element by node numbers, the numbers shall be counterclockwise, as is shown in Fig. 5.62.

The positive direction of normal $n$ to the shell element is defined by the right hand rule regarding the nodal sequence of element. The upper surface of the shell is defined as the positive side of the normal. When element 352 in the detail B of

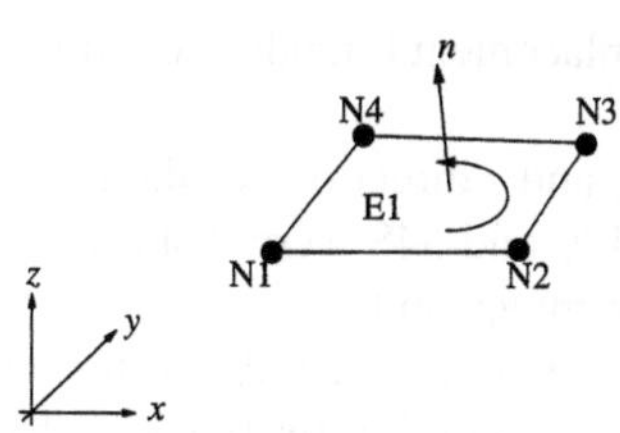


Figure 5.62: A shell element and its constituting nodes.

Table 5.29: Boundary condition (Analytical model 1)

| node NO. | displacement (mm) | | | rotation (rad.) | | |
|---|---|---|---|---|---|---|
| | $x$ | $y$ | $z$ | $x$-axis | $y$-axis | $z$-axis |
| 162 | 0.0 | 0.0 | 0.0 | free | free | free |
| 533 | 0.0 | 0.0 | 0.0 | free | free | free |
| 340 | 0.0 | 0.0 | 0.0 | free | free | free |
| 178 | 0.0 | 0.0 | 0.0 | free | free | free |
| 534 | 0.0 | 0.0 | 0.0 | free | free | free |
| 356 | 0.0 | 0.0 | 0.0 | free | free | free |
| 164 | 0.0 | 0.0 | 0.0 | free | free | free |
| 516 | 0.0 | 0.0 | 0.0 | free | free | free |
| 342 | 0.0 | 0.0 | 0.0 | free | free | free |
| 170 | −19.7 | −57.4 | 0.0 | free | free | free |
| 348 | −19.7 | −57.4 | 0.0 | free | free | free |

Table 5.30: Boundary condition (Analytical model 2)

| node NO. | displacement (mm) | | | rotation (rad.) | | |
|---|---|---|---|---|---|---|
| | $x$ | $y$ | $z$ | $x$-axis | $y$-axis | $z$-axis |
| 162 | 0.0 | 0.0 | 0.0 | free | free | free |
| 533 | 0.0 | 0.0 | 0.0 | free | free | free |
| 340 | 0.0 | 0.0 | 0.0 | free | free | free |
| 178 | 0.0 | 0.0 | 0.0 | free | free | free |
| 534 | 0.0 | 0.0 | 0.0 | free | free | free |
| 356 | 0.0 | 0.0 | 0.0 | free | free | free |
| 164 | 0.0 | 0.0 | 0.0 | free | free | free |
| 516 | 0.0 | 0.0 | 0.0 | free | free | free |
| 342 | 0.0 | 0.0 | 0.0 | free | free | free |
| 1001 | 0.0 | 0.0 | 0.0 | 0.0 | 0.0 | 1.571 |
| 2001 | 0.0 | 0.0 | 0.0 | 0.0 | 0.0 | 1.571 |

Fig. 5.60 is observed, the constituting nodes are described in the sequence of 533, 534, 356, and 340. The surface of the visible side of the element is the upper side of the shell. For the material data, Young's modulus equals 205.8 GPa and Poisson's ratio 0.315.

Table 5.29 shows the displacement boundary condition for analytical model 1 in Fig. 5.60.

The displacement in $x$, $y$, and $z$ directions of the node which is fixed to the shaft were restrained and nodes 170 and 348 were displaced in $x$ and $y$ directions by an amount equivalent to the pin movement.

Table 5.30 is displacement boundary condition for analytical model 2 shown in Fig. 5.61. While the boundary condition for the node fixed to the shaft is the same as analytical model 1, the rotation of the pin is expressed by rotating nodes 1001 and 2001 about $z$-axis. The degree-of-freedom for 1001 and 2001 other than this rotation is restrained.

**(2) Analytical result and accuracy** Fig. 5.63 and Fig. 5.64 show the deformed shape of analytical model 1.

It is clear from the figures that the deformed shape is not the one that could be obtained by rotating the outer end around the center of the shaft. Analytical model 1 is not appropriate.

Fig. 5.65 and Fig. 5.66 show the deformed shape of analytical model 2. Although the deformed shape obtained by this model looks appropriate, which could be achieved by rotating the outer end, the elongated deformation of truss element is unrealistically too large. To proceed further, geometrical nonlinear analysis was carried out using the same geometry of analytical model 2. The results are shown in Fig. 5.67 and Fig. 5.68.

The figures show that the appropriate deformed shape was obtained by geometrical nonlinear analysis. Hence, the geometrical nonlinearity analysis is essential to obtain the exact shape of the spiral spring after deformation.

Next, we will examine the characteristics of the spring.

Fig. 5.69 shows the analytical result of spring characteristics.

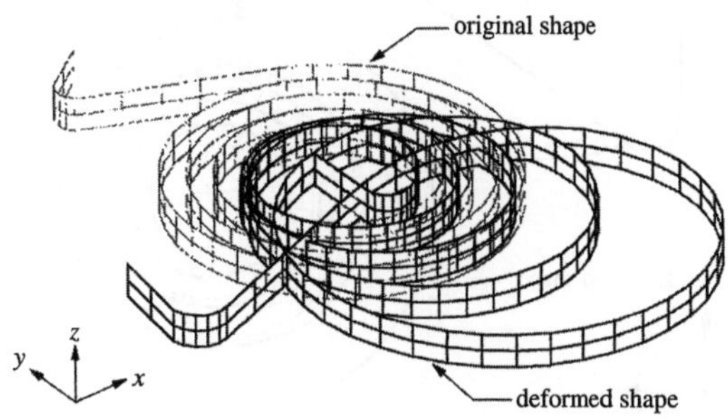


Figure 5.63: Deformed shape of analytical model 1.

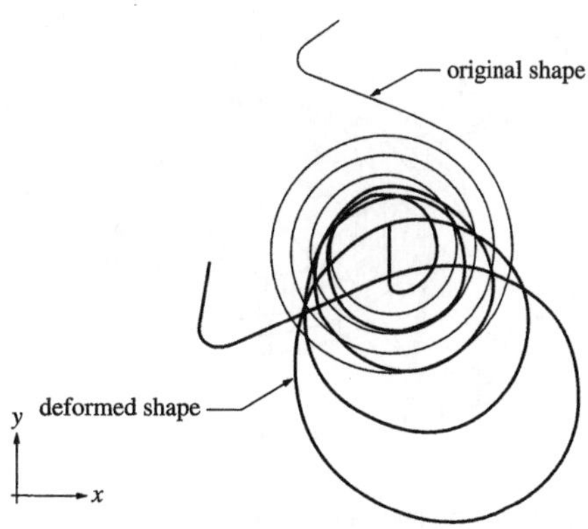


Figure 5.64: Deformed shape of analytical model 1.

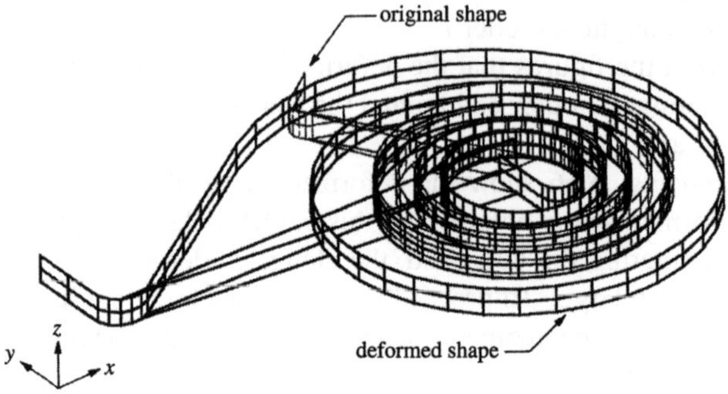


Figure 5.65: Deformed shape of analytical model 2.

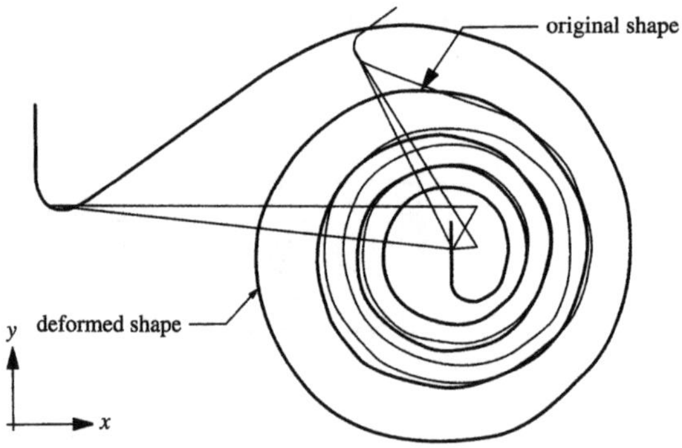


Figure 5.66: Deformed shape of analytical model 2.

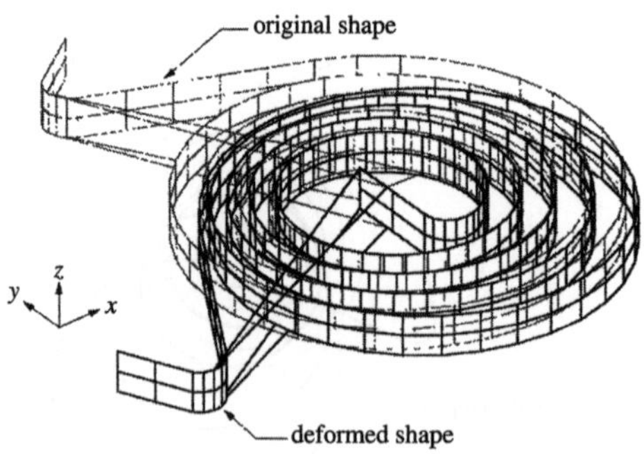


Figure 5.67: Deformed shape of analytical model 2 by geometrical nonlinear analysis. (Three-dimensional view)

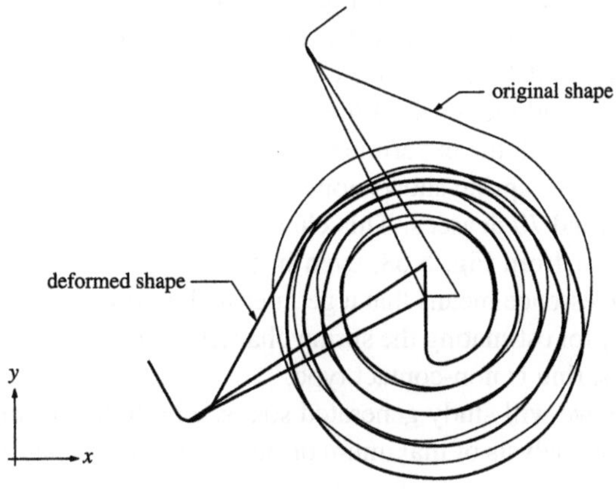


Figure 5.68: Deformed shape of analytical model 2 by geometrical nonlinear analysis. (Plane View)

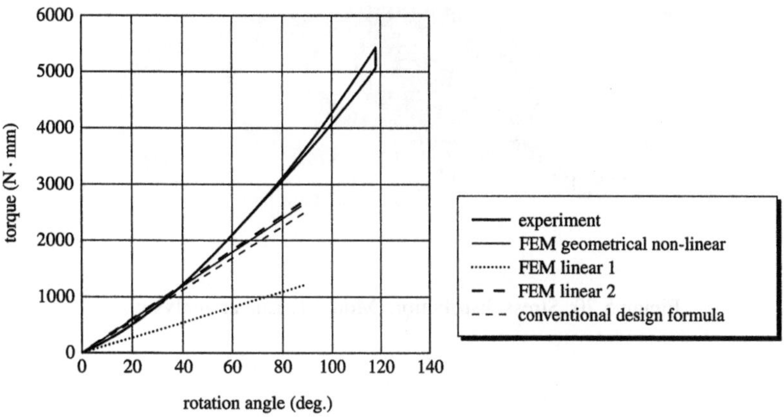


Figure 5.69: Spring characteristics.

The abscissa of the figure is rotation angle of the outer end around center shaft and the ordinate is torque. The result obtained by the conventional formula (5.12)[25] is shown here together with the FEM analysis explained in this paragraph.

$$K = \frac{EI}{1.25L}, \tag{5.12}$$

where $E$ is Young's modulus and $I$ is the moment of inertia of area. $L$ is the length of the spiral spring. The developed length (= 522 mm) from point a to b in Fig. 5.58 was used here.

We can observe from the figure that the result of the linear analysis using analytical model 1 is quite different from the experimental value. This difference arises from the fact that the analyzed motion of the outer end of the spring rotating about the center of the shaft does not reflect actual move, as we see from the deformed shape shown in Fig. 5.64. The result of the linear and the geometrical nonlinear analyses in model 2 and the result of the conventional design formula show a good degree of agreement with the experimental value, to roughly 40 degrees of wind-up angle. Accordingly, the spring shows nonlinearity over 40 degrees of wind-up angle and hysteresis beyond 70 degrees. This characteristic was caused by the contact of the coils, as is clear from Fig. 5.68, even in the case of the non-contact type spiral spring. This in essence means that a geometrical nonlinear analysis with friction factor is necessary for estimating the spring characteristic with any hope of accuracy even if the spiral spring is non-contact type.

As a final step, we will study generated stress. Fig. 5.70 and Fig. 5.71 show the results of the linear analysis of maximum principal stress by analytical models 1 and 2, respectively.

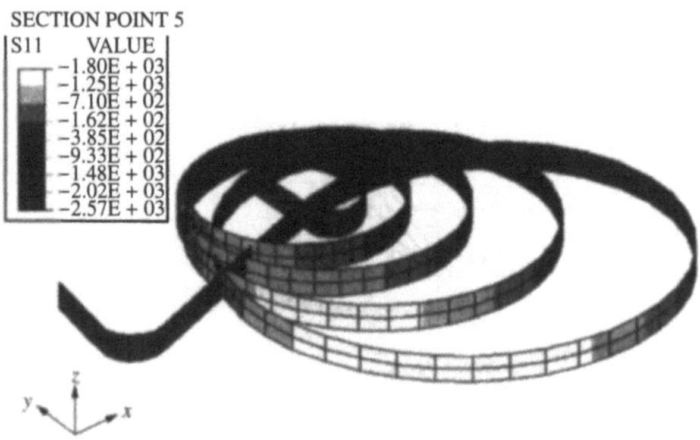


Figure 5.70: Stress distribution (Model 1: Linear analysis).

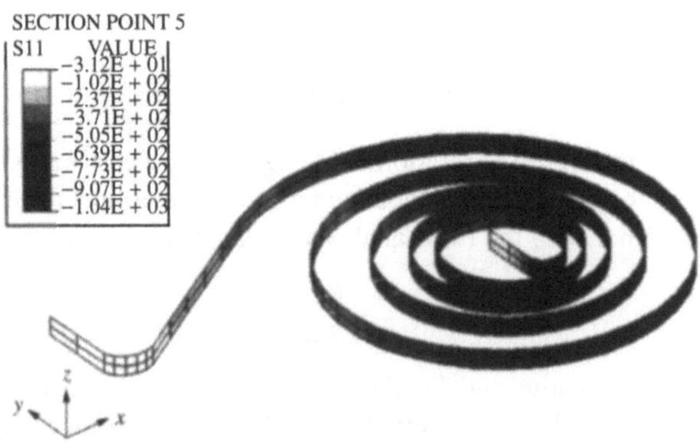


Figure 5.71: Stress distribution (Model 2: Linear analysis).

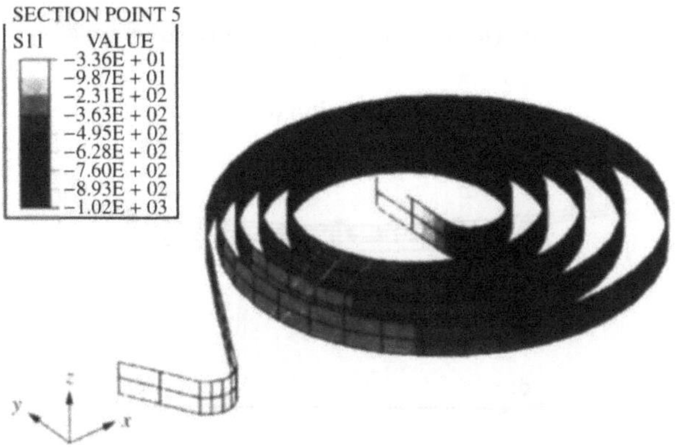


Figure 5.72: Stress distribution (Model 2: Nonlinear analysis).

Fig. 5.72 shows the result of the geometrical nonlinear analysis according to analytical model 2.

The peak values of the maximum principal stress are 2.59 GPa based on the linear analysis of analytical model 1, 1.04 GPa based on the linear analysis of analytical model 2 and 1.03 GPa based on the geometrical nonlinear analysis of analytical model 2 respectively. On the other hand, stress by conventional design formula (5.13) is 0.98 GPa.

$$\sigma = \frac{12M_0}{bh^2}, \tag{5.13}$$

where $b$ is the width of strip, $h$ is the thickness of strip, and $M_0$ is the torque applied to the shaft.

It could be said that the peak values of the maximum principal stress by the linear analysis and the geometrical nonlinear analysis of analytical model 2 are nearly equal to that of the conventional design formula. Analytical model 1 is twice as large as the results shown for model 2. Furthermore, the position of the maximum stress varies distinctly in the models between Fig. 5.71 and Fig. 5.72 when we compare both closely. So, the influence of the geometrical nonlinearity cannot be avoided even in the stress analysis.

## 5.7 LEAF SPRINGS

The leaf spring is widely used for the suspension systems of automobiles and railroad cars. It has an advantage to serve as a structural member in the suspension mechanism together with a spring function.

Typical types of the leaf spring are shown in Fig. 5.73.

Type (a) is the most widely used standard type of spring and it is called the

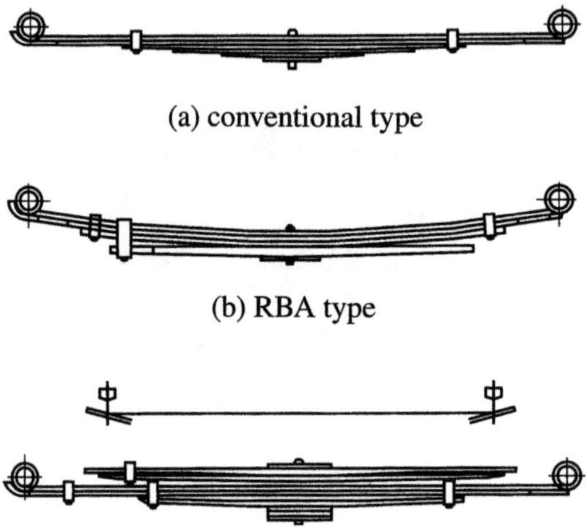

(a) conventional type

(b) RBA type

(c) main & helper type

Figure 5.73: Typical leaf springs.

conventional type leaf spring. Its spring characteristic is almost linear within small deflection.

Type (b) is called the progressive leaf spring or RBA type leaf spring (Rolling Bearing Auxiliary Spring). With this type of spring, the main spring and the auxiliary spring begin to contact and the contact point moves along with the load showing smooth nonlinearity in the spring characteristic.

Type (c) is called the two-stage progressive spring. This type of spring has a sharp nonlinear characteristic. The helper spring begins to work when it contacts the bearing block with the main spring being deflected to a certain amount.

In the analysis of spring types (b) or (c), it is important to consider the change of contact condition between the main spring and the auxiliary spring or between the auxiliary spring and the bearing block. In other words, an analysis by the contact element is necessary. The analysis of this type of spring will be studied in chapter 6. The conventional type is going to be discussed in this chapter. Since the contact between leaves is induced even in the type (a) spring, the analysis by the contact element[26] is necessary when a rigorous result is desired. In this paragraph, however, it is necessary to express the contact condition by an assumption for simplification without using contact element. In the case of the conventional design formula (tip contact method; see Paragraph 3.1.1), three assumptions are laid down.

(a) The contact between spring leaves occurs at the leaf tip and center.

(b) The spring leaf is considered to be a straight beam.

(c) The deformation is small.

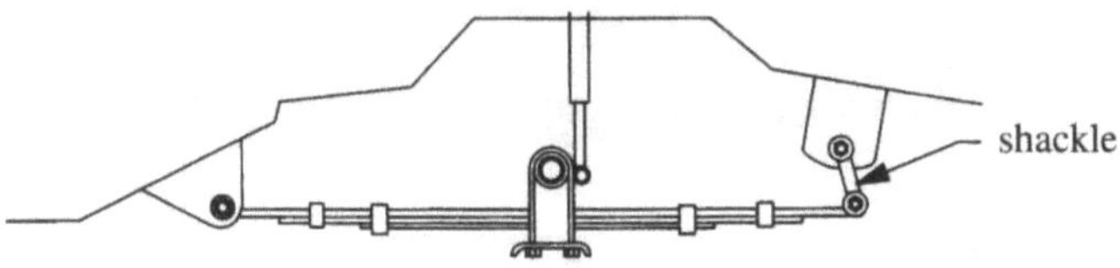


Figure 5.74: Installed state of a leaf spring.

Let us assess the tip contact method[27] by performing linear analysis on supposition (a) and (c). The result will be checked against experiment.

When a leaf spring is used for the suspension of vehicles, it is installed onto the vehicle by a shackle, as is shown in Fig. 5.74.

The shackle will considerably affect the spring characteristic. Therefore, an analysis that includes the shackle is necessary in designing a leaf spring. It is also necessary to analyze the effect of horizontal force, wind-up torque and vertical load, because the spring receives those forces when a vehicle brakes or accelerates. In addition, the stress known as nip stress occurs within spring when fastened by a center bolt or a buckle. This stress has a noticeable effect on the durability of spring. So we also have to include the effect of nip clamping in our analysis. An analysis that includes the shackle or the action of wind-up torque will be discussed in chapter 6. As to the fastening with nip by a center bolt, an evaluation will be discussed here.

**(1) Modeling and input data** Two design formulas have been accepted for a leaf spring. One is the developed leaf method or equi-curvature method and the other is the tip contact method. The former is a calculating method working under the assumption that overlapping surfaces are contacting each other in their full length and the latter assumes that the transmission of force from one leaf to the next leaf is carried out only through tips. As an improved version, the modified tip contact method[28] proposes that there exists a contact point at the tip of the spring leaf No.$i + 2$ as well as at the tip of the spring leaf No.$i + 1$, as is shown in Fig. 5.75, between spring leaf No.$i$ and spring leaf No.$i + 1$.

We will proceed on the latter concept. The geometry and the dimensions of an objective leaf spring are shown in Fig. 5.76 and Table 5.31.

The leaf is numbered in the order of length in Table 5.31. $R$ is the radius of leaf curvature. Fig. 5.77 shows an outline of the analytical model.

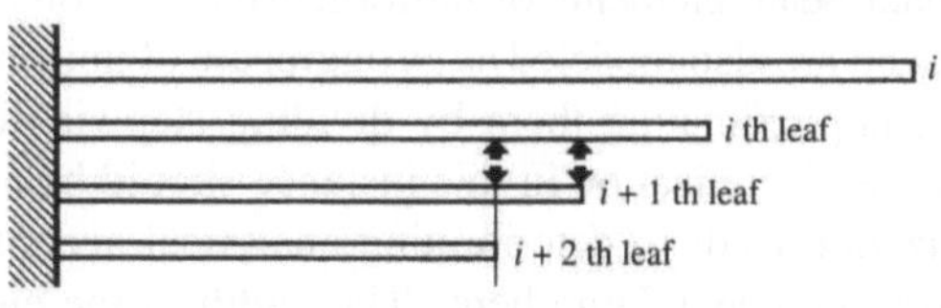


Figure 5.75: Contact points in modified tip contact method.

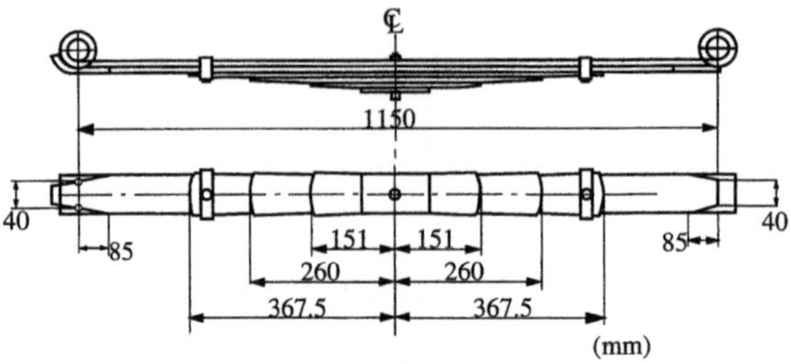


Figure 5.76: Geometry and dimension.

Table 5.31: Dimensions of leaves.

| leaf NO. | width (mm) | thickness (mm) | $R$ (mm) | $a$ (mm) | $b$ (mm) | 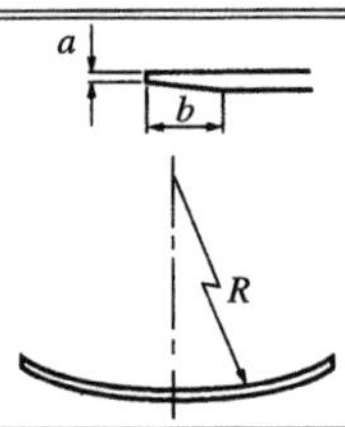 |
|---|---|---|---|---|---|---|
| 1 | 69.8 | 10.0 | 2048 | – | – | |
| 2 | 69.8 | 10.0 | 1825 | – | – | |
| 3 | 70.3 | 10.2 | 2153 | 5.0 | 110.0 | |
| 4 | 70.0 | 10.2 | 3332 | 5.0 | 110.0 | |
| 5 | 70.0 | 9.2 | 1195 | 4.5 | 110.0 | |

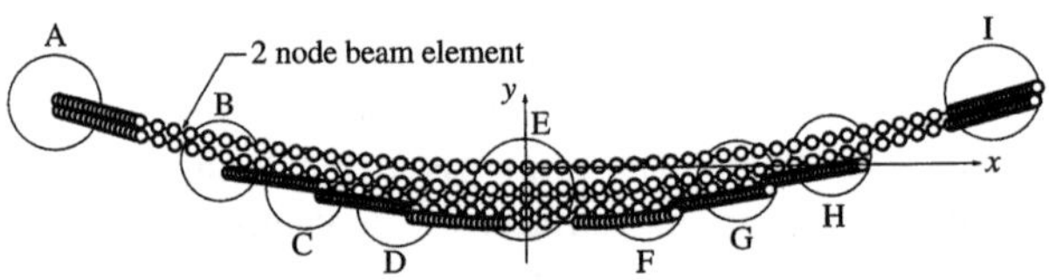


Figure 5.77: Outline of analytical model.

Circles in the figure show node positions. These nodes are located on the curve running through the leaf thickness center in the bundled condition. Their coordinate values are expressed by $x$-$y$ coordinate system shown in Fig. 5.77.

Under the condition of a concentrated load at the center of the leaf spring, the two-dimensional element is sufficient for our analysis since our only concern is the deformation within $x$-$y$ plane. Hence, the spring leaf was modeled by two-dimensional two-nodal beam elements of uniform cross-section. The tapered section and the triangle end are also modeled as an aggregate of uniform cross-sectional beam elements, i.e., approximating them by dividing stepwise into a number of strips. The mesh of the discretization in this instance should be as fine as possible since rough discretization leads to deteriorating analytical accuracy. The element length was assumed to be about 5 mm here. The width of the element was represented by the midpoint of element. The length of the element was assumed to be 20 mm, except for the fine discretization of 5 mm mentioned above.

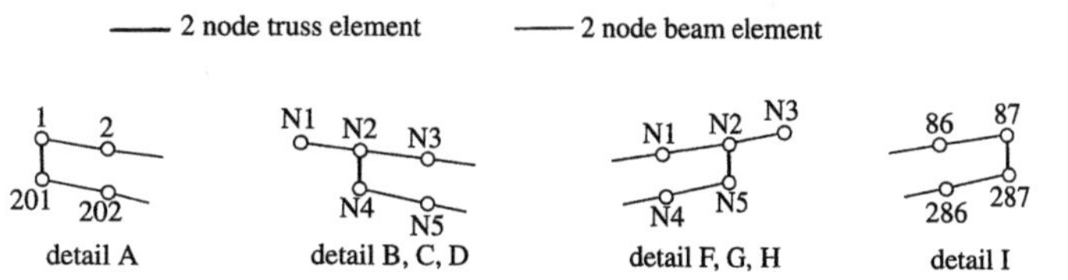


Figure 5.78: Models of interleaf contact (tip contact method).

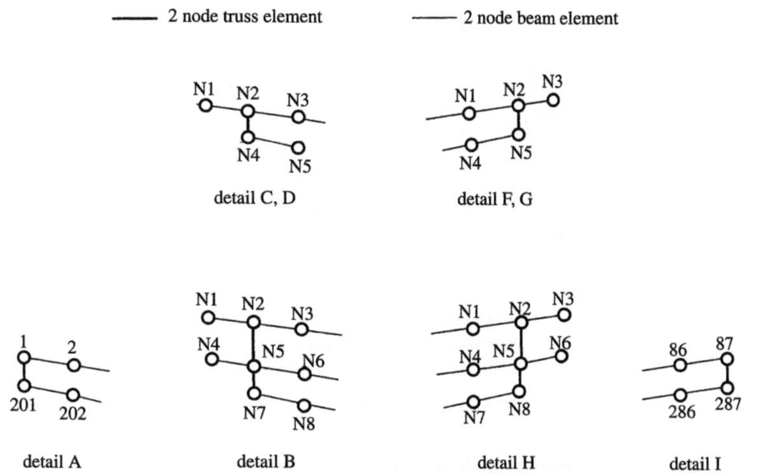


Figure 5.79: Models of interleaf contact (modified tip contact method).

The details of area A to D and area F to I of Fig. 5.77 are shown in Fig. 5.78 and Fig. 5.79.

Since only the compressive force transmits at tip locations, the transmitting condition of the force has been modeled by a two-nodal truss element. The elements denoted by the bold line in Fig. 5.78 and Fig. 5.79 are the two-nodal truss element. The concept of the modified tip contact method[28] was applied only to the contact between No.1 and No.2 leaves in Fig. 5.79. The numbers of nodes near the interleaf contact points of the model are shown in Fig. 5.78 and Fig. 5.79. To save space, some nodes are expressed by the notations N1 to N8 instead of numerals. The correspondence between this notation and the actual number of the node is shown in Table 5.32 and Table 5.33.

Nodes in the shadowed area of the tables are truss elements. The detail E of Fig. 5.77 is shown in Fig. 5.80.

Model (a) is used for the analysis in which the concentrated load is applied to the center of spring. Model (b) is used for analysis of stress (nip stress) when the

Table 5.32: Nodal number at interleaf contact (tip contact method).

| detail | N1 | N2 | N3 | N4 | N5 |
|---|---|---|---|---|---|
| B | 225 | 226 | 227 | 401 | 402 |
| C | 422 | 423 | 424 | 601 | 602 |
| D | 621 | 622 | 623 | 801 | 802 |
| F | 636 | 637 | 638 | 847 | 848 |
| G | 446 | 447 | 448 | 658 | 659 |
| H | 261 | 262 | 263 | 468 | 469 |

Table 5.33: Nodal numbers at interleaf contact (modified tip contact method).

| detail | N1 | N2 | N3 | N4 | N5 | N6 | N7 | N8 |
|---|---|---|---|---|---|---|---|---|
| B | 25 | 26 | 27 | 225 | 226 | 227 | 401 | 402 |
| C | 422 | 423 | 424 | 601 | 602 | – | – | – |
| D | 621 | 622 | 623 | 801 | 802 | – | – | – |
| F | 636 | 637 | 638 | 847 | 848 | – | – | – |
| G | 446 | 447 | 448 | 658 | 659 | – | – | – |
| H | 61 | 62 | 63 | 261 | 262 | 263 | 468 | 469 |

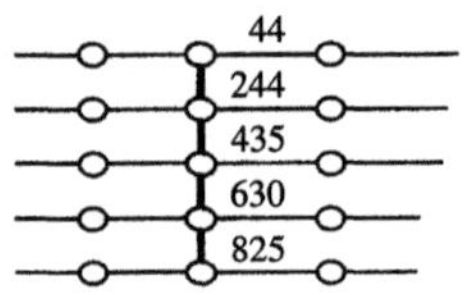


(a) When a concentrated load is applied at the center of spring

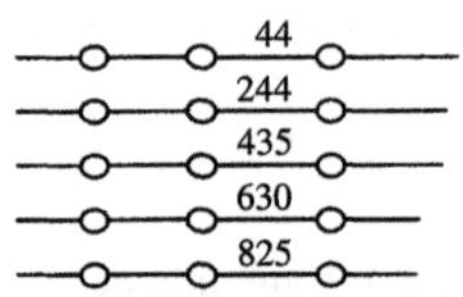


(b) When the spring is fastened by a center bolt

Figure 5.80: Models of the center section (Detail E).

center bolt is fastened. When the concentrated load is applied at the center of the spring after fastening by the center bolt, stress can be obtained by superposing these two kinds of stresses. In model (a), center bolt coupling condition is expressed by connecting the nodes at center on each spring leaf with two-nodal beam elements. The geometry of spring leaf after fastening should be used in the modeling since the shape of each spring leaf differs in the occasion before and after the fastening. For convenience, the present analysis is based on the shape before the fastening.

In model (b), the fastening by the center bolt is expressed by the forced deflection of nodes at the center of spring leaf.

Table 5.34: Nodal coordinate

| node NO. | x (mm) | y (mm) |
|---|---|---|
| 1 | −564.48 | 80.17 |
| 44 | 0.00 | 0.00 |
| 87 | 567.48 | 80.17 |
| 201 | −565.02 | 69.29 |
| 244 | 0.00 | −22.94 |
| 287 | 565.02 | 69.29 |
| 401 | −366.21 | −1.71 |
| 435 | 0.00 | −33.11 |
| 469 | 365.23 | −1.92 |
| 601 | −258.80 | −27.65 |
| 630 | 0.00 | −49.28 |
| 659 | 257.81 | −27.77 |
| 801 | −152.10 | −51.50 |
| 825 | 0.00 | −61.07 |
| 848 | 148.62 | −51.84 |

Table 5.35: Data of beam element

| element NO. | node NO. | | width (mm) | thickness (mm) |
|---|---|---|---|---|
| 1 | 1 | 2 | 69.81 | 10.20 |
| 43 | 43 | 44 | 69.81 | 10.20 |
| 86 | 86 | 87 | 69.81 | 10.20 |
| 201 | 201 | 202 | 69.83 | 10.15 |
| 243 | 243 | 244 | 69.83 | 10.15 |
| 286 | 286 | 287 | 69.83 | 10.15 |
| 401 | 401 | 402 | 70.23 | 5.46 |
| 434 | 434 | 435 | 70.28 | 10.19 |
| 468 | 468 | 469 | 70.23 | 4.93 |
| 601 | 601 | 602 | 70.02 | 10.14 |
| 629 | 629 | 630 | 70.02 | 10.14 |
| 658 | 658 | 659 | 70.02 | 5.47 |
| 801 | 801 | 802 | 70.11 | 9.17 |
| 824 | 824 | 825 | 70.11 | 9.07 |
| 847 | 847 | 848 | 70.11 | 4.71 |

A portion of the nodal coordinate and the beam element data used for the model is shown in Table 5.34 and Table 5.35.

The size shown in the table is the measured value. The element coordinate system $(t, n_1, n_2)$ of the beam element is shown in Fig. 5.81.

The direction of vector $n_1$ and $n_2$ shall be specified. Table 5.36 is the data of truss element representing the contact condition between spring leaves.

The truss elements No.9113 and No.9114 in the table are the only elements nec-

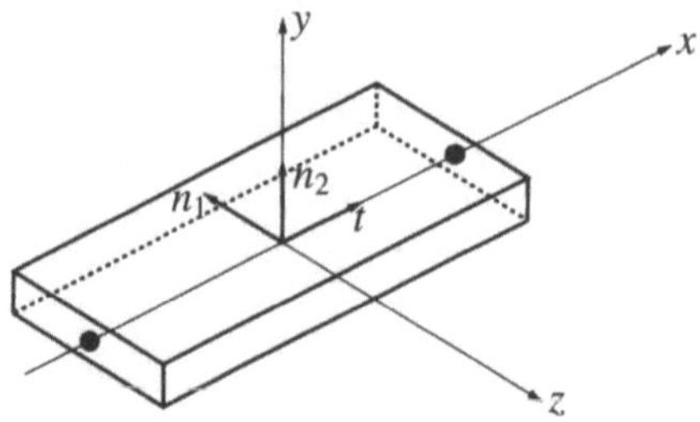


Figure 5.81: Element coordinate system of beam element $(t, n_1, n_2)$.

Table 5.36: Data of truss element

| element NO. | node NO. | | cross-sectional area ($mm^2$) |
|---|---|---|---|
| 9101 | 1 | 201 | 100 |
| 9102 | 87 | 287 | 100 |
| 9103 | 226 | 401 | 100 |
| 9113 | 226 | 26 | 100 |
| 9104 | 262 | 469 | 100 |
| 9114 | 262 | 62 | 100 |
| 9105 | 423 | 601 | 100 |
| 9106 | 447 | 659 | 100 |
| 9107 | 622 | 801 | 100 |
| 9108 | 637 | 848 | 100 |

Table 5.37: Data of beam element.

| element NO. | node NO. | | shape & radius (mm) |
|---|---|---|---|
| 9210 | 44 | 244 | circle: $R5.0$ |
| 9211 | 244 | 435 | circle: $R5.0$ |
| 9212 | 435 | 630 | circle: $R5.0$ |
| 9213 | 630 | 825 | circle: $R5.0$ |

essary for the model of modified tip contact method. The data representing Fig. 5.80(a) is shown in Table 5.37.

The element coordinate system $(t, n_1, n_2)$ of the beam element in Table 5.37 is shown in Fig. 5.82.

The cross-sectional area of truss element and the radius of beam element were chosen adequately.

As for the material data, Young's modulus was assumed to be 195 GPa and Poisson's ratio to be 0.3, for all elements. Young's modulus is measured value for spring leaves.

To analyze the stress and the deformation of a leaf spring with a concentrated load being applied at its center, the nodes at the eyes (nodal numbers 1 and 87 in

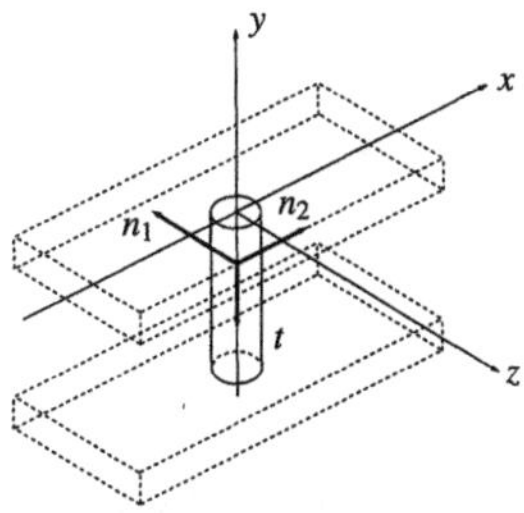


Figure 5.82: Element coordinate system of beam element.

Table 5.38: Boundary condition under the concentrated load.

| node NO. | displacement (mm) | | rotation (deg.) |
|---|---|---|---|
| | $x$ | $y$ | $z$-axis |
| 1 | free | 0.0 | |
| 87 | | | |
| 44 | 0.0 | | |
| 244 | | | free |
| 435 | | free | |
| 630 | | | |
| 825 | | | |

Fig. 5.79) are supported as a simple beam and the designated load is applied at node (nodal number 825) of No. 5 leaf in the positive direction of $y$ axis. The eyes moves freely in $x$ direction when the load is applied. The node (nodal number 44) at the center bolt (see Fig. 5.80) shall be restrained to prevent any rigid body movement in the direction of $x$. In the same manner, nodal number 244, 435, 630, and 825 shall also be restrained in $x$ direction. The boundary condition for this is shown in Table 5.38.

The applied load is assumed to be 16 kN.

The boundary condition for the analysis of the nip stress by the model of Fig. 5.80(b) is as follows. The fastening condition by the center bolt is shown in Fig. 5.83. Fastening by center bolt is simulated by the forced displacement given to each leaves.

$v_i$ in Fig. 5.83 is the amount of the displacement of No.$i$ spring leaf and can be obtained by the following formula:

$$v_i = \sum_{j=2}^{i} \left\{ (y_{j-1} - y_j) - \frac{(t_{j-1} + t_j)}{2} \right\}, \tag{5.14}$$

where $y_j$ is $y$ coordinate of the node at the center bolt of No.$j$ spring leaf and $t_j$ is the thickness of No.$j$ spring leaf. Resultant boundary condition of the fastening is shown in Table 5.39.

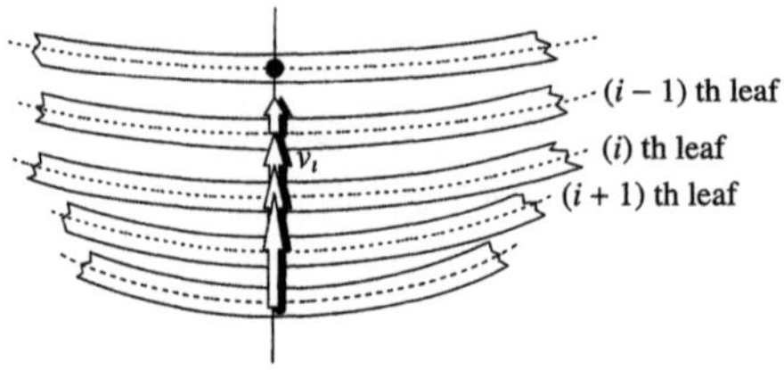


Figure 5.83: Fastening by a center bolt.

Table 5.39: Boundary condition of fastening.

| node NO. | displacement (mm) | | rotation (deg.) |
|---|---|---|---|
| | $x$ | $y$ | $z$-axis |
| 44 | 0.0 | 0.0 | 0.0 |
| 244 | | 12.765 | free |
| 435 | | 12.765 | |
| 630 | | 18.770 | |
| 825 | | 20.905 | |

**(2) Analytical result and accuracy** Firstly, the validity of tip contact method and modified tip contact method will be checked. Fig. 5.84 shows the deformed shape when a concentrated load is applied to the center of the leaf spring in tip contact method.

Fig. 5.85 is a similar result by modified tip contact method.

The solid line in the figure is the shape after the deformation and the dotted line is the shape before the deformation. It is clear from Fig. 5.84 that the surfaces of the spring leaves No.1 and No.2 perforate each other at area A in the case of tip contact method. We see an inequality as follows:

$$y_1(x) - y_2(x) < \frac{t_1(x) + t_2(x)}{2},$$

where $y_i(x)$ is the $y$ position of $i$-th leaf at the location of $x$ (see Fig. 5.77) and $t_i$ is the thickness of $i$-th leaf at the location of $x$. As far as Fig. 5.85 is concerned,

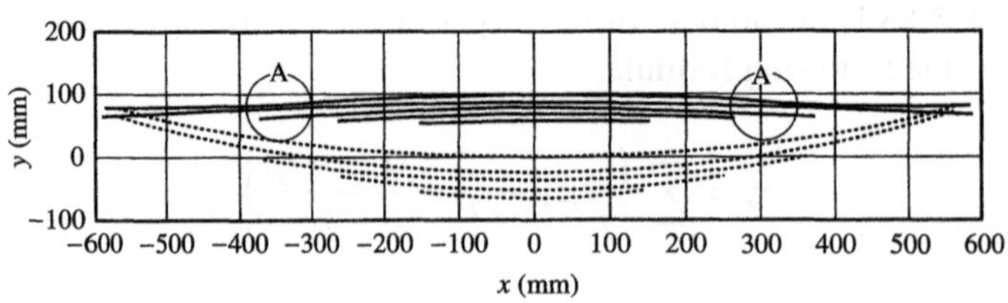


Figure 5.84: Deformed shape of the model by tip contact method.

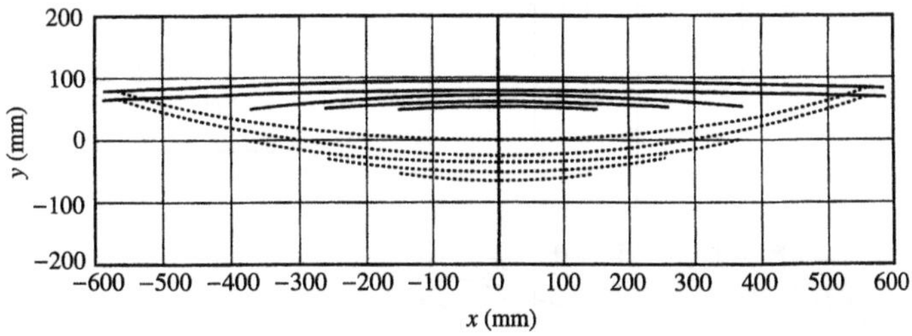


Figure 5.85: Deformed shape of model by the modified tip contact method.

such an area cannot be observed using the modified tip contact method. A detailed investigation is still needed to judge if the inequality shown above exists in the area or not.

In any case, assumption[29] of tip contact method or modified tip contact method is only valid when the span or the thickness of the leaves satisfies a certain condition and modified tip contact method seems more appropriate than tip contact method in the present case.

Fig. 5.86 and Fig. 5.87 show the analytical results of stress by each method.

The stress shown in the figure is the value at the integral point since the element stiffness was obtained by numerical integration in this model. The abscissa of the figure shows distance $x$ measured from the center of the spring before the deformation to the integral point of beam element and the ordinate shows stress $\sigma$. The experimental value for the spring leaves No.1 and No.2 is shown by the circular marks in each figure. Nip stress is not factored into the experimental or the analytical value in the figure. As is apparent from the experimental results, the stress of spring No.1 is almost equal to that of No.2 within the range of $x = \pm 300$ mm and variation of the stress along the spring leaf is very small. In opposition to this, the stress obtained by tip contact method, differs significantly between spring leaves

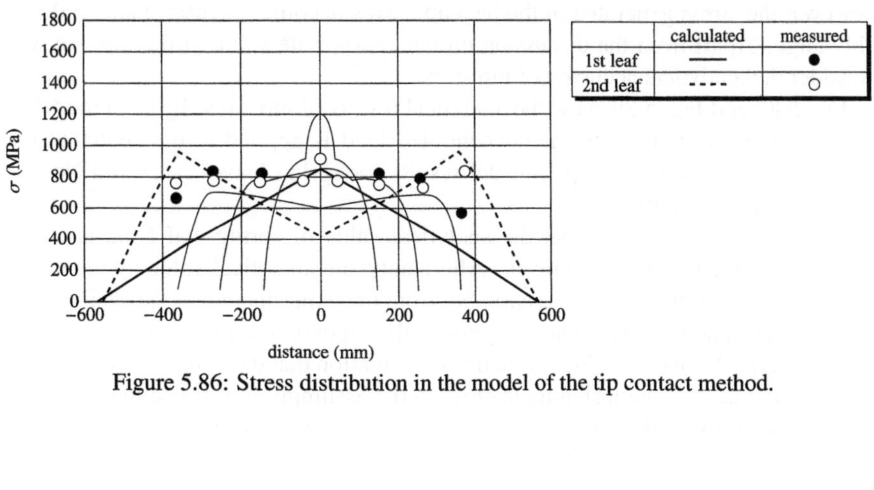


Figure 5.86: Stress distribution in the model of the tip contact method.

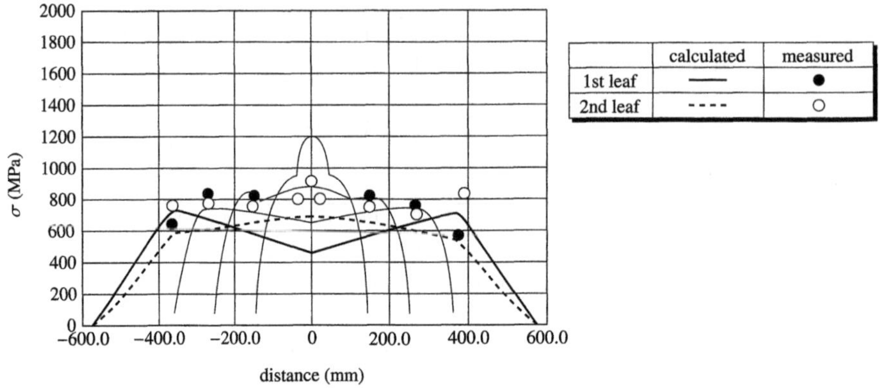


Figure 5.87: Stress distribution in the model of the modified tip contact method.

No.1 and No.2 and variation of stress along spring leaf is substantial.

It can be seen from Fig. 5.87 that the stress of spring leaves No.1 and No.2 by modified tip contact method does not coincide with the experimental value. However, the shape of the stress distribution along the spring leaf approaches the experimental value due to the increase of contact points between leaves. This shows that the stress distribution in the leaf spring changes considerably by the condition of contact between leaves. In other words, a precise description of the contact condition between leaves is essential to the stress calculation of leaf spring when high accuracy is desired.

Now let's consider the leaf surface on which positive or tensile stress is induced when concentrated load is applied. When the spring is fastened, negative stress is induced on the upper side of the main leaf and some adjacent leaves, while the positive stress is induced on the backup leaves. Therefore, nip stress contributes to the improvement of the durability of the main leaf. When nip is given too much, however, the stress generated in the backup leaves becomes excessive and undesirably large. Apart from the discussion on the appropriate value of the nip, we will consider the analytical accuracy of nip stress.

Fig. 5.88 and Fig. 5.89 show the analytical results of nip stress by each model.

$\sigma$ shall be positive when the concentrated load is applied at the center of the spring. In the figure, the experimental results for spring leaves No.1 and No.2 is shown by circular marks.

In Fig. 5.88 we observe that the analytical value of spring leaf No.2 is closer than No.1 to the experimental result. Alternately, the analytical value by the modified tip contact method differs remarkably from the experimental value, as is shown in Fig. 5.89. The nip stress varies significantly depending on the contact state between leaves. We may conclude from this examination that the contact state between leaves in the case of the fastening is close to the assumption of tip contact method. Precise modeling of the contact state between the leaves will be necessary for the

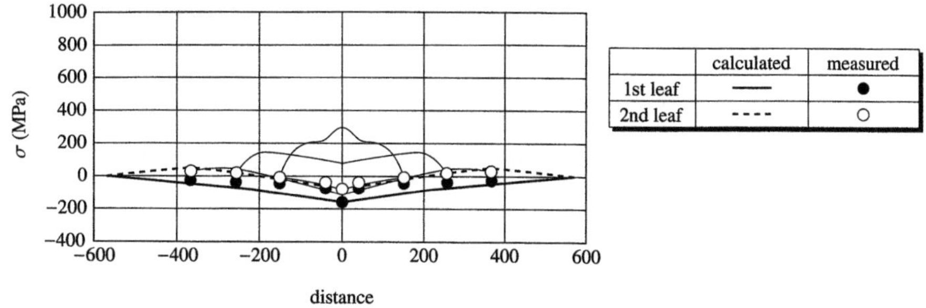


Figure 5.88: Nip stress distribution (tip contact method)

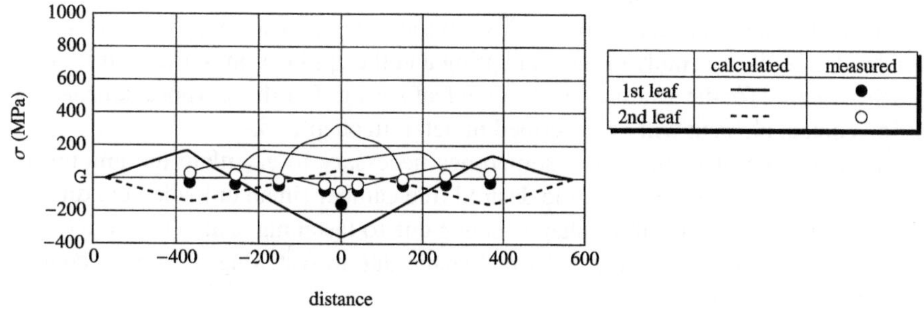


Figure 5.89: Nip stress distribution (modified tip contact method)

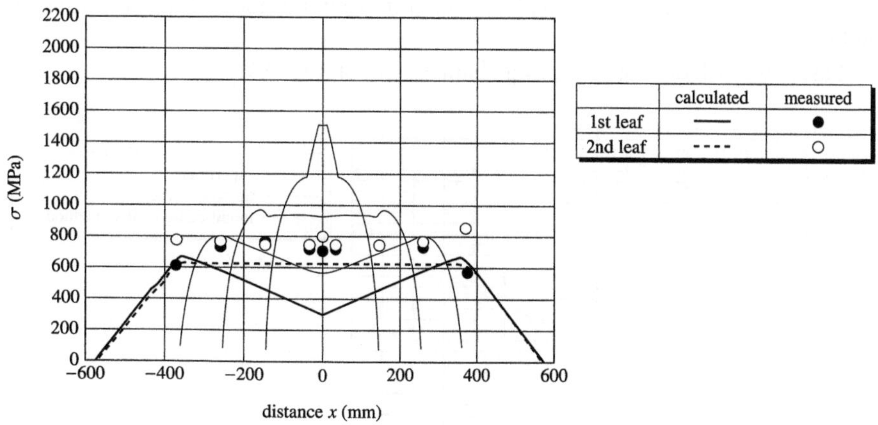


Figure 5.90: Stress distribution taking the nip stress into account.

further improvement of analytical accuracy.

The stress under the concentrated load and fastening force (nip stress) can be obtained by superposing. The result is shown in Fig. 5.90.

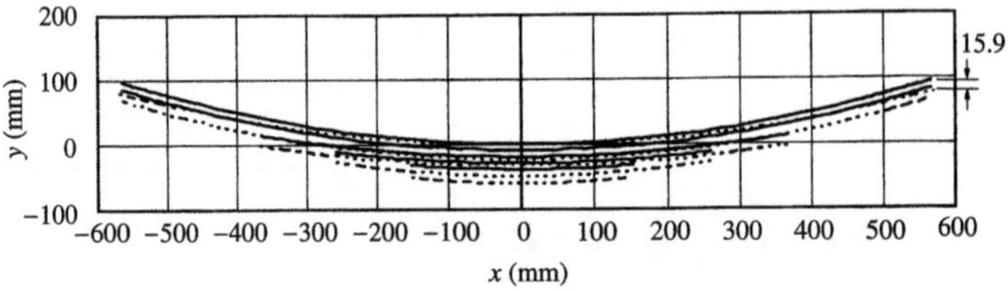


Figure 5.91: Deformed shape by fastening (a model by tip contact method).

In this figure, stress due to the concentrated load is represented by modified tip contact method shown in Fig. 5.87 and nip stress is represented by tip contact method shown in Fig. 5.88.

It may be clear now that a precise description of the contact state between spring leaves during the assembly of the leaf spring and the application of the load is absolutely essential for the accuracy of stress in leaf spring. For this purpose, an analysis using contact element will be described in detail in chapter 6.

In the design of springs, it is sometimes necessary to calculate the amount of fastening deformation to assure the desired free camber (in an unloaded condition). To this end, the analysis of the shape change due to fastening is also necessary.

The analytical result of the deformed shape due to fastening by the tip contact method is shown in Fig. 5.91.

Solid line in the figure shows the shape after fastening and dotted line is the original shape. Both ends of the spring leaf No.1 are displaced by 15.9 mm upwards by the fastening. Although we have no experimental results to compare, the analysis seems to be reasonable in practical use, judging from the result of nip stress shown in Fig. 5.88.

Lastly, we will discuss the spring characteristic in Fig. 5.92.

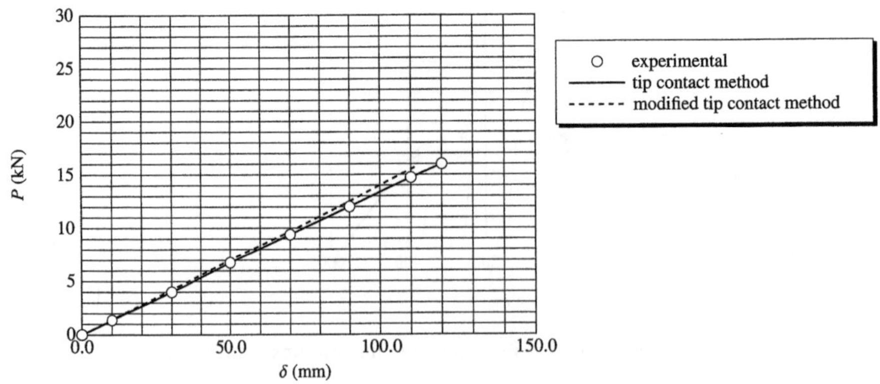


Figure 5.92: Relation between load and deflection.

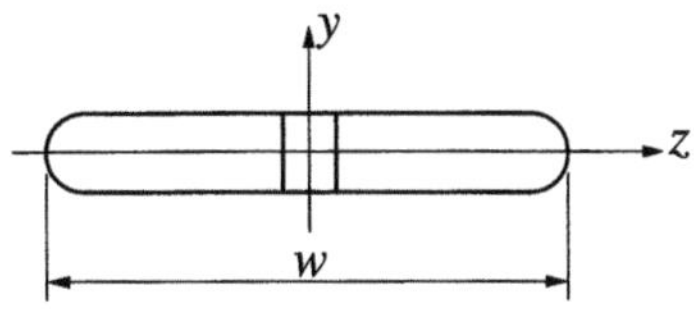


Figure 5.93: Decrease of cross-sectional area compensating rounded edge and center hole.

Solid line in the figure shows the results using tip contact method and broken line shows results by modified tip contact method. The experimental results are shown with circular marks. The analysis by tip contact method is in close accord with the experimental value with a 0.9% error, while the spring constant (load $P/$deflection $\delta$) by modified tip contact method shows about a 4.3% higher value than the experimental value. In any case, the spring characteristic can be analyzed accurately enough to be of practical use.

Generally, a spring leaf has a cross section with rounded side-edges and a center bolt hole at the center as shown in Fig. 5.93.

It is necessary to consider the decrease of cross-sectional area when the spring leaf is modeled by the beam element of a rectangular cross-section. This can be done by modifying dimension $w$ so that the moment of inertia of area about $z$-axis has the same value as the decreased cross-sectional area. In addition to the precise modeling of contact state between the leaves, the modification of cross-sectional dimension is an important consideration for the improvement of analytical accuracy. The spring analyzed here will decrease by about 2 % if the cross-sectional area is modified taking the roundness of side-edge into account. Geometrical nonlinearity is an equally important consideration for the improvement of analytical accuracy. The spring constant will decrease a further 1 % if geometrical nonlinearity is accounted for in this case.

## 5.8 FLAT SPRINGS

A spring that is made of sheet metal by a press machine is generically referred to as a flat spring. The flat spring is used in many fields as a mechanical component for small equipment and a variety of shapes are available. Hence, in the design of the flat spring, a calculating formula that can be adapted to each shape and the state of usage is required. There are many flat springs that have holes or notches, designs which take the stress concentration into account are indispensable for such springs. The design formula included in a handbook on springs, however, is only for the simple geometry and it is necessary to develop an adequate formula in person for the spring with complex geometry. To create a design formula capable of both functional accuracy and stress concentration is quite difficult. Traditionally, this spring evaluation depends on an experiment. The flexibility of geometric conditions

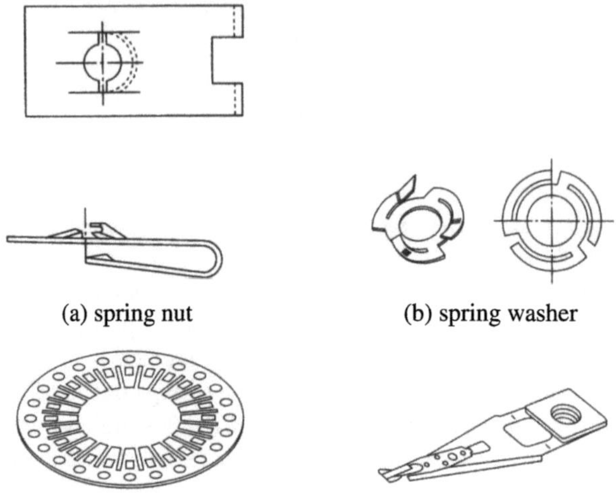

(a) spring nut (b) spring washer

(c) strike wire drive spring for dot printer (d) HDD suspension

Figure 5.94: Examples of flat springs.

is a big advantage of FEM.

Some representative examples of the flat spring are shown in Fig. 5.94. In addition to the variety of shapes, functional performance varies significantly.

In the case of a suspension spring for the magnetic head of hard disc drives, the evaluation of the frequency of resonance[30] is important to prevent a tracking error (writing or reading error) due to the vibration of spring. An analysis of the natural frequency and the natural mode of vibration will be discussed in this section.

**(1) Analytical model** The spring was modeled by the shell element.

The outline of the analytical model is shown in Fig. 5.95.

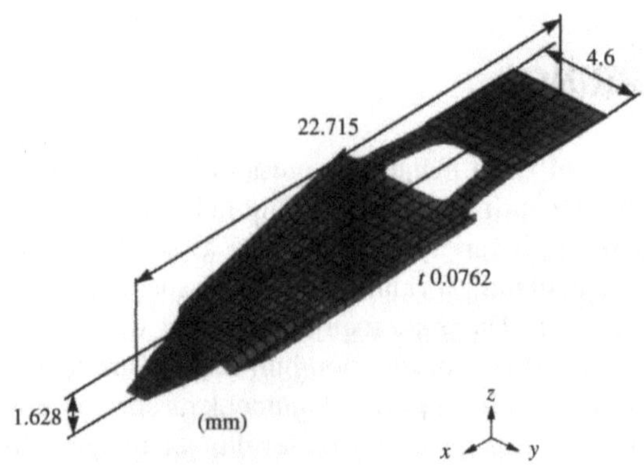


Figure 5.95: An analytical model.

A typical dimension of the spring is also shown in the figure. The data of nodes (coordinate values of the nodes), the elements (kind of the elements, constituting nodes, and thickness of the elements), the material characteristic (Young's modulus, Poisson's ratio), and the boundary condition are necessary. The natural frequency can be analyzed by specifying the mass density of material and the maximum order of vibration in addition to the above stated data. (To avoid redundancy, those data will not appear.)

An analysis up to the tenth order was obtained on assumptions of a homogeneous material with a boundary condition restraining six-degree-of-freedom of mount base (shown as rectangular shape in Fig. 5.95).

**(2) Analytical result** The analytical result is shown in Fig. 5.96.

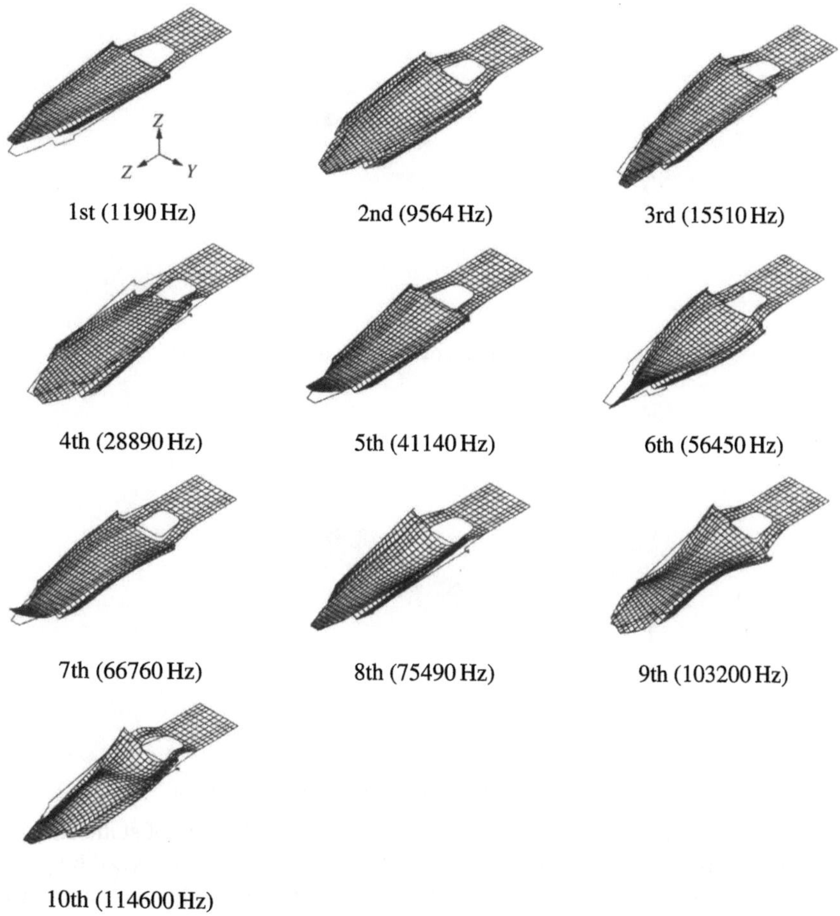


Figure 5.96: Natural frequency and natural mode of vibration.

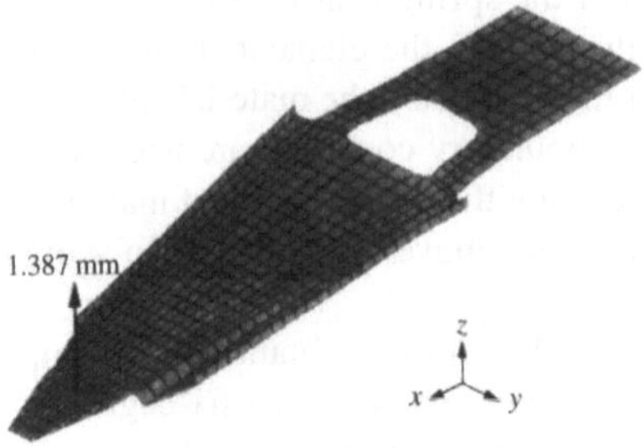


Figure 5.97: Condition of forced displacement.

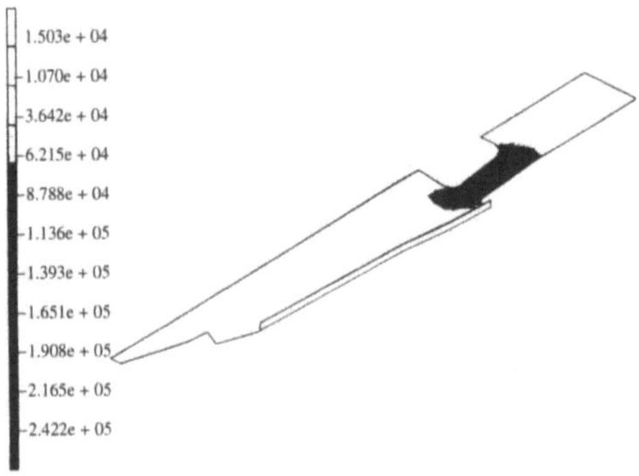


Figure 5.98: Distribution of maximum principal stress.

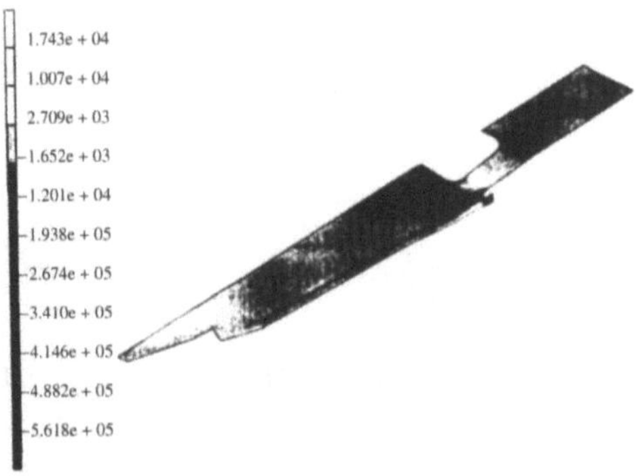


Figure 5.99: Distribution of minimum principal stress.

This analysis has been carried out simply by replacing data out of static analysis. A reverse analysis is also possible. Stress or deformation can be analyzed by giving a load condition or a forced displacement condition instead of the maximum order of vibration. As an example, the distribution of maximum principal stress and minimum principal stress when a node near the tip is given the forced displacement of 1.387 mm, as is shown in Fig. 5.97, are displayed in Fig. 5.98 and Fig. 5.99.

## 5.9 STRESS CONCENTRATION

In springs, there is stress concentration in several places, such as, round holes, notches, and areas of sharp change in the cross-sectional area. For improved design analysis, the accurate evaluation of stress concentration is essential. FEM demonstrates its formidable powers of calculation in cases of such stress concentration.

In this section, some stress concentration analysis for the periphery of the center bolt hole, the slit bottom of disc springs, and the terminal of torsion bars, as is shown in Fig. 5.100, will be discussed with some tips and points on the analysis.

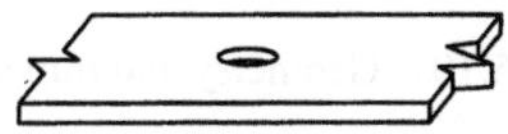

(a) bolt hole periphery in spring leaf

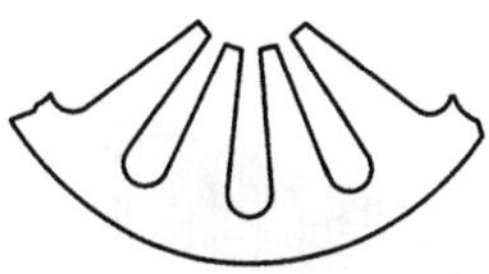

(b) slit bottom in disc spring

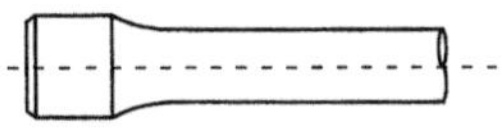

(c) grip end of torsion bar

Figure 5.100: Examples of stress concentration in springs.

### 5.9.1 Stress Concentration on the Periphery of a Center Bolt Hole for Leaf Springs

Usually, the central part of leaf spring is in a tightly clamped state as it is installed on an axle shaft with a U-bolt. As a result of this, stress around central part is not exceptionally high. By way of comparison, some springs may be installed on the axle shaft with an inserted rubber sheet (iso-clamping). In some cases, U-bolt might loosen in service. The stress concentration around center bolt hole may cause delayed fracture or fatigue fracture in such cases.

**(1) Analytical model and the point of modeling** An example of the stress concentration around a hole when bending moment is applied will be presented. A detailed report[31] is published concerning stress concentration when bending moment is applied to a flat member with a central round hole. Here, we will analyze the stress concentration of a case in which the tension-side corner of the hole is rounded off as is shown in Fig. 5.101.

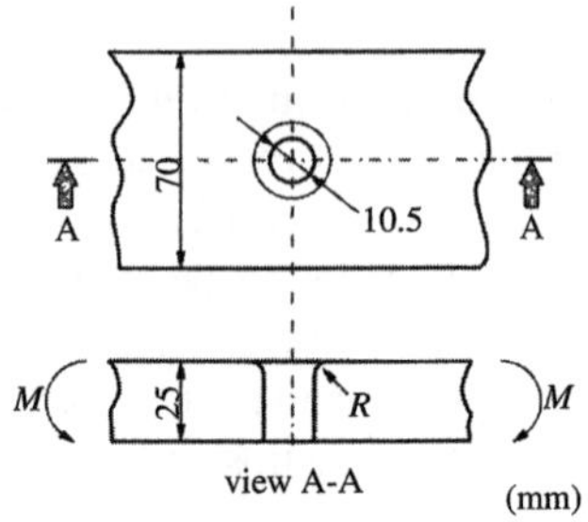


Figure 5.101: Geometry and dimension.

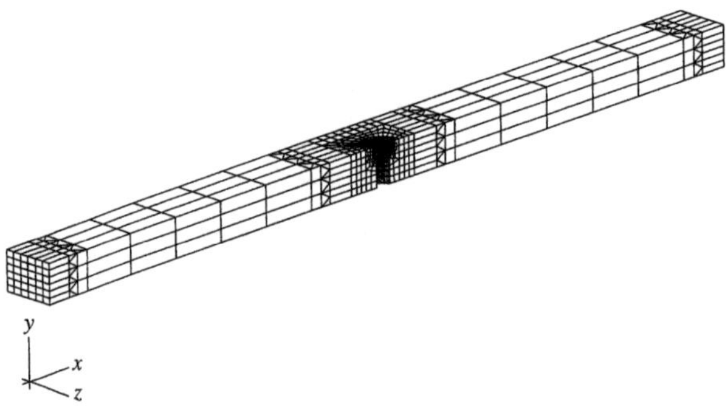


Figure 5.102: Analytical model.

The measurements of the analytical spring leaf is, as is shown in Fig. 5.101, 70 mm in width, 25 mm in thickness, and 10.5 mm in the diameter of the hole. The corner radius $R$ of the hole was evaluated in four levels:1.5 mm, 2.5 mm, 5.0 mm, and 7.5 mm. In the preparation of an analytical model, it is necessary to set the length of the spring leaf so that the disturbance of stress around loaded point does not affect stress distribution around center bolt hole. In this regard, the length of the spring leaf was set at 550 mm. Fig. 5.102 shows the analytical model with a 2.5 mm $R$.

The discretization around the hole on the compression side of spring leaf is coarse since our main concern is the concentration of stress on tension side.

The elements used in the analysis are solid elements with 20 nodes and 15 nodes. The necessary input data is not different from those explained in other analytical examples, namely, nodal coordinate value, the kind of element used, the constituting node of element, material characteristic and boundary condition. For the solid element, the constituting nodes are described in the order of N1, N2, …, N20 or N1, N2, …, N15, as is shown in Fig. 5.103.

When we make a model by solid elements, the number of nodes or elements will

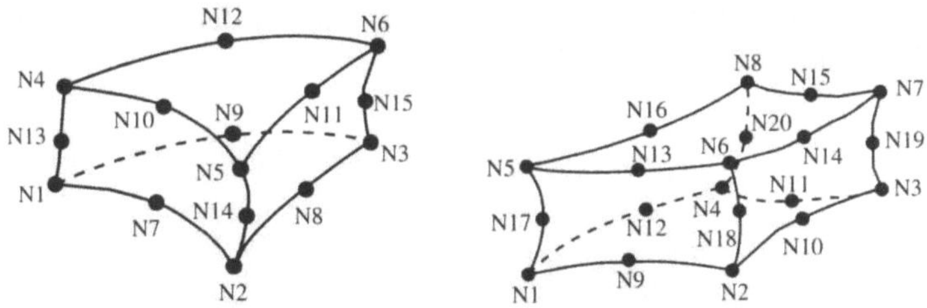


15 node solid element 20 node solid element

Figure 5.103: Solid elements.

Table 5.40: Boundary condition.

| | displacement (mm) | | |
|---|---|---|---|
| | $x$ | $y$ | $z$ |
| nodes on surface | free | free | free |
| nodes on A-A | 0.0 | 0.0 | free |
| nodes on B-B | free | 0.0 | free |

naturally increase. When we analyze stress concentration, in particular, a fine discretization of concerned area is necessary and the model size becomes considerably large. As a consequence, it takes more time to run computer. To carry out this analysis effectively, expertise is necessary to construct a compact model using the symmetry of deformation or shape. For our present analysis, a half part was modeled as is shown in Fig. 5.102 and the boundary condition for nodes on cut plane was specified taking the symmetry of deformation into account. The boundary condition is shown in Table 5.40.

A-A and B-B shown in the table means the neutral axis on the end sections of spring leaf shown in Fig. 5.104.

For the load condition, couples of forces are given to nodes constituting upper and lower surfaces at both end sections as shown in Fig. 5.104. The magnitude of the couple was specified so that the bending moment acting on spring leaf equals 1.312 kN·m.

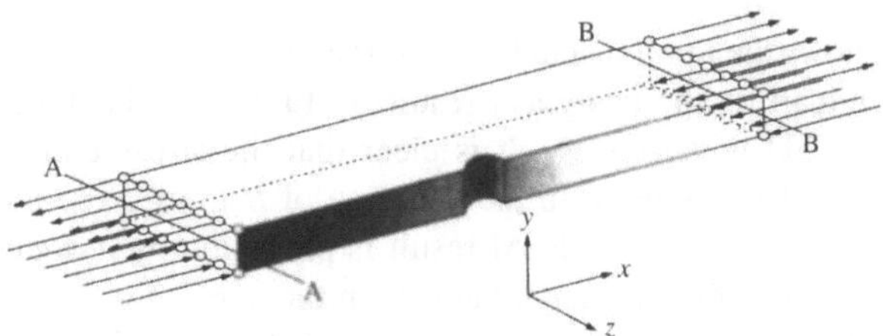


Figure 5.104: Boundary condition and load condition.

**(2) Analytical result** Fig. 5.105 is a stress contour diagram in the neighborhood of the hole.

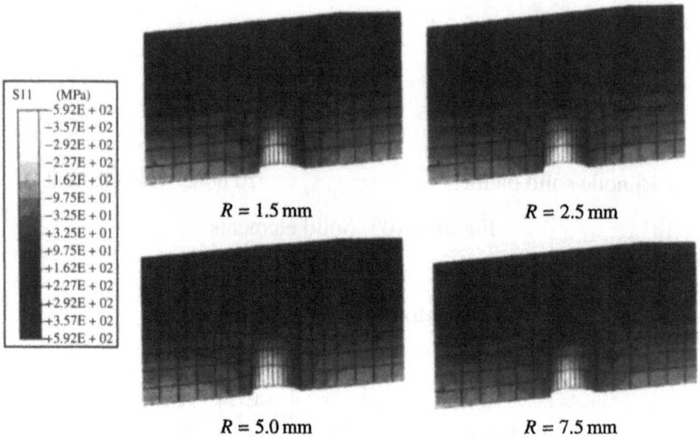


Figure 5.105: Stress distribution in the neighborhood of hole.

The finite element mesh is also superposed in this diagram. The discretization around hole corner goes more finely with the decrease of corner radius, because the stress concentration factor increases as the radius of rounded edge decreases. As is apparent from the diagram, the stress reaches a maximum value at the position just inside the surface. We define stress concentration factor $\alpha$ using the following formula to express the degree of concentrated stress.

$$\alpha = \frac{\sigma_{max}}{\sigma_0}, \tag{5.15}$$

$$\sigma_0 = \frac{6M}{(b-d)t^2}, \tag{5.16}$$

where $\sigma_{max}$ is the maximum stress generated on the periphery of hole when bending moment $M$ is applied. $b$, $t$, and $d$ are width, thickness and hole diameter respectively.

Then, the effect of radius $R$ of hole corner on stress concentration factor will be discussed.

Fig. 5.106 shows the relationship between $R$ and $\alpha$.

The circular mark is the analytical result by FEM and the square mark is the theoretical value[31] with zero R. It is clear that the stress concentration factor $\alpha$ obtained by FEM increases with the decrease of $R$ toward the theoretical value shown by the square mark. The FEM result is presented by a dotted line. In this case, $\alpha$ for 1.5 mm $R$ shows a smaller value than the line.

We can attribute this to insufficient fineness of discretization at the area where stress concentration took place. To analyze the stress concentration accurately and effectively, it is necessary to properly estimate stress distribution to some degree before starting analysis. Then, to repeatedly conduct the discretization and the anal-

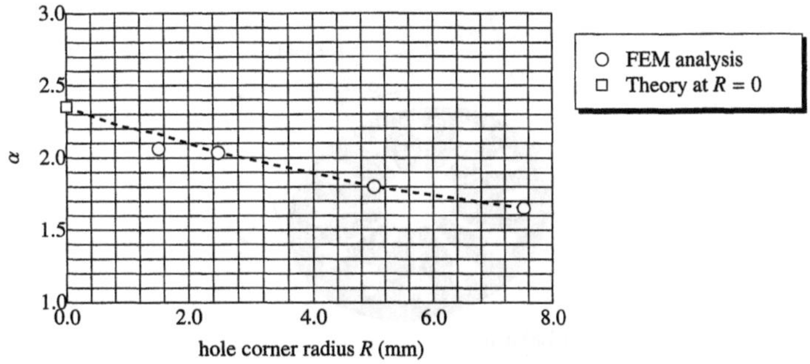


Figure 5.106: Stress concentration factor.

ysis until the solution converges to a certain value, while, at the same time deciding whether or not the discretization is fine enough.

Incidentally, there is now commercial software available that can automatically carry out troublesome calculations that would otherwise need constant reiteration. Difficult tasks such as analyzing the stress concentration, can now be accomplished by even a novice designer. In FEM software equipped with such a function, the method used is called adaptive mesh generation[32]. It automatically changes discretization so that the analytical error becomes quite small.

For the readers information, there are three methods for modifying discretization, these are the $h$ method, $r$ method and the $p$ method. The $r$ method changes the coordinates of the nodes without changing the number of elements. The $h$ method divides the element into a finer mesh where larger error is observed. In $p$ method, the order of the interpolating function of the element in the area where the error is large will be increased to a higher order.

### 5.9.2 Stress Concentration at the Slit Bottom of a Disc Spring

Almen-Laszlo's equation or Wempner's equation are both used as general design formula for disc springs[33]. These equations are only approximate equations derived from the following assumptions.

(1) The rotating displacement of each minute region (shell element) of the disc spring is small.

(2) The conical shape is kept after the spring is deflected.

(3) The influence of meridian strain is neglected.

In the case of the disc spring with slits as shown in Fig. 5.107, it is a common way to carry out an approximate calculation on an assumption that cantilevers are joined to form a disc spring.

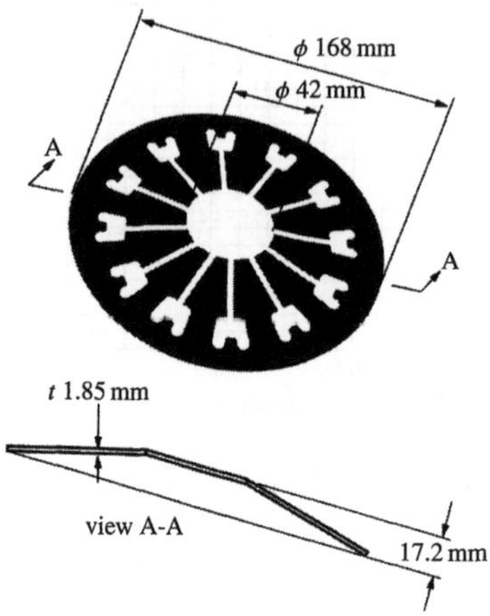


Figure 5.107: Geometry of disc spring.

It is essential to pay attention to the shape of the slit bottom to lessen the stress concentration, but the conventional design formula mentioned above is not capable of handling this situation.

**(1) Analytical model** The stress distribution of the disc spring shown in Fig. 5.107 is analyzed here. The approximate dimension of the disc spring shown in Fig. 5.108 is the analytical model. Only one twenty-fourth of the segments is modeled making use of the symmetry of shape and deformation.

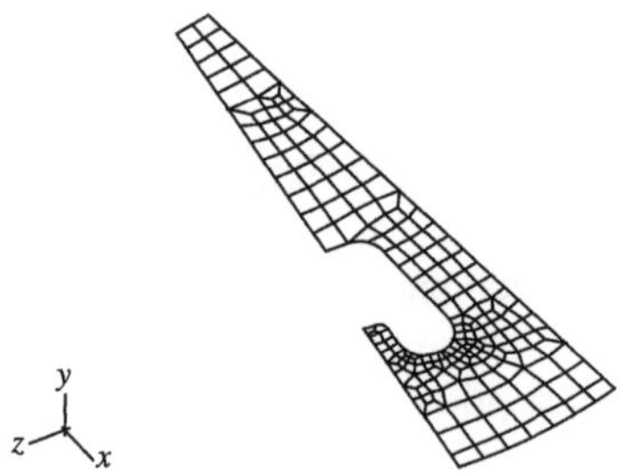


Figure 5.108: An analytical model.

The analytical model was constructed using eight-nodal shell elements. The material characteristic was assumed to be homogeneous, Young's modulus was assumed to be 205.8 GPa and Poisson's ratio 0.315.

The analysis was carried out on the condition that displacement in $y$ direction of nodes on arc a-a shown in Fig. 5.109 is restrained and the nodes on arc d-d are displaced by 7.0 mm downwards.

Nodes on lines b-b and c-c shown in Fig. 5.109 give the boundary condition

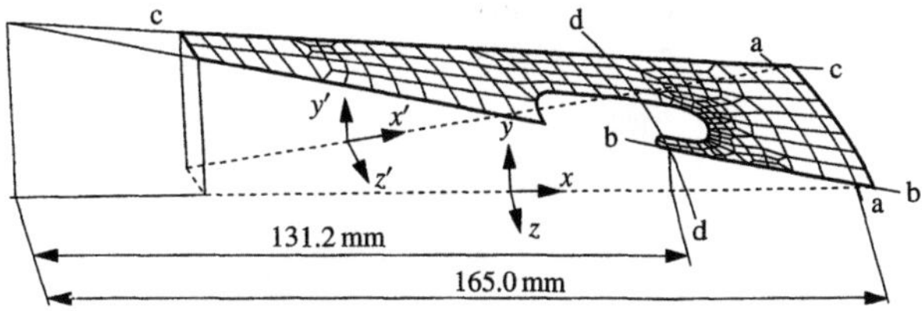


Figure 5.109: Boundary condition and load condition.

Table 5.41: Boundary condition of displacement.

| node | displacement (mm) | | | rotation (deg.) | | |
|---|---|---|---|---|---|---|
| | $x$, $x'$ | $y$, $y'$ | $z$, $z'$ | $x$-axis, $x'$-axis | $y$-axis, $y'$-axis | $z$-axis, $z'$-axis |
| on arc a-a | free | 0.0 | free | free | free | free |
| on line b-b | free | free | 0.0 | 0.0 | 0.0 | free |
| on line c-c | free | free | 0.0 | 0.0 | 0.0 | free |
| on arc d-d | free | 7.0 | free | free | free | free |

representing the symmetry of the deformation. The detailed boundary condition of displacement is in Table 5.41.

**(2) Analytical result** The relation between deflection and load obtained by linear analysis and geometrical nonlinear analysis is shown in Fig. 5.110.

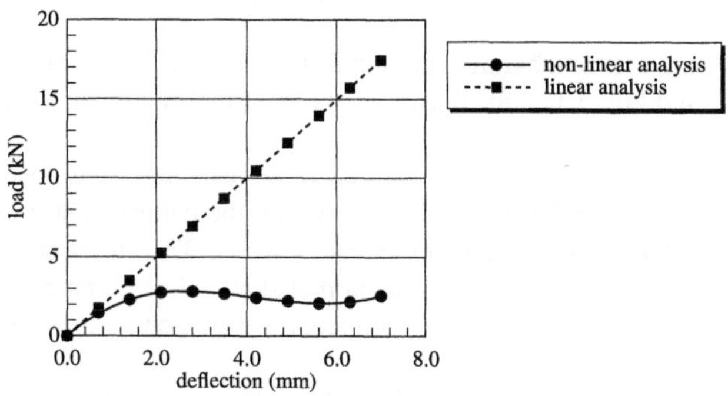


Figure 5.110: Relation between load and deflection.

The abscissa shows deflection (the absolute value of $y$ displacement) of nodes on arc d-d and the ordinate shows reaction force in $y$ direction (24 times as large as the partial result) acting on each node on arc a-a shown in Fig. 5.109. The circular marks in the figure are the result of geometrical nonlinear analysis and the square marks are those of linear analysis. As is evident in the figure, a jump phenomenon peculiar to disc spring is observed in geometrical nonlinear analysis, but not in linear analysis.

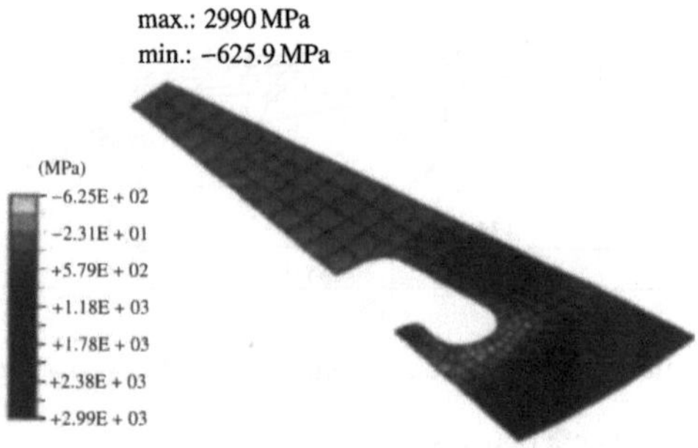


(a) back side

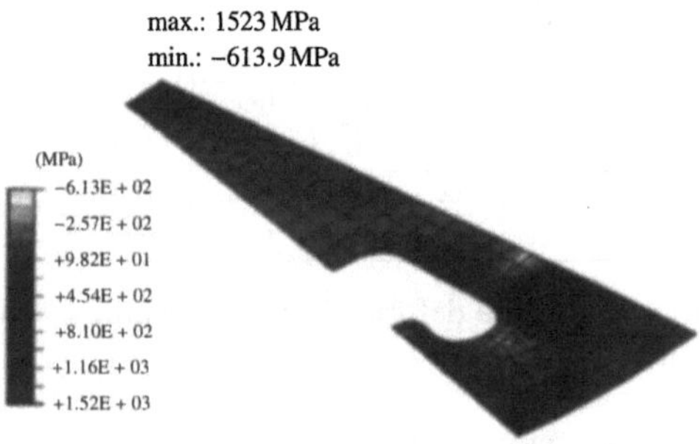


(b) top side

Figure 5.111: Distribution of maximum principal stress (linear analysis).

The distribution of the maximum principal stress at a deflection of 7 mm is shown in Fig. 5.111 and Fig. 5.112 for linear and geometrical nonlinear analysis respectively.

It is clear from these figures that no large stress concentration is observed around the slit part of this particular disc spring. However, the pattern of stress distribution differs between linear analysis and nonlinear analysis. This difference is attributed to buckling which causes a jump phenomenon of shell. The influence of geometrical nonlinearity cannot be ignored in stress concentration calculation at the slit bottom of disc spring.

On a related note, residual stress that takes place within springs during manufacturing process through press forming, shot peening, or presetting, should be included in the evaluation of stress concentration. However, the precise estimation of stress distribution cannot be accomplished simply by superposing residual stress, because the residual stress generated during manufacturing process affects load deflection characteristic, especially jump phenomena. When we want to estimate stress distribution precisely, advanced analysis is necessary, including the simulation of manufacturing process, the analysis under operating conditions and so on.

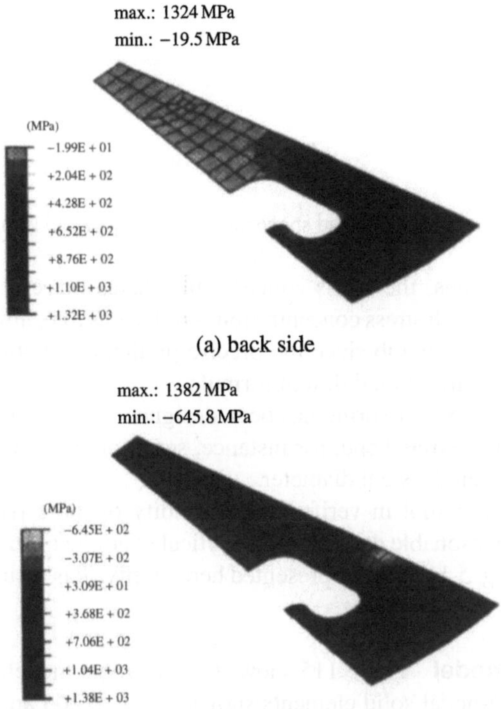


Figure 5.112: Distribution of maximum principal stress (nonlinear analysis).

### 5.9.3 Stress Concentration at the End of a Torsion Bar Spring

A spring in the shape of a straight bar which utilizes twisting resilience is called a torsion bar spring. For convenience in installation, the ends of a torsion bar spring are formed into shapes as is shown in Fig. 5.113.

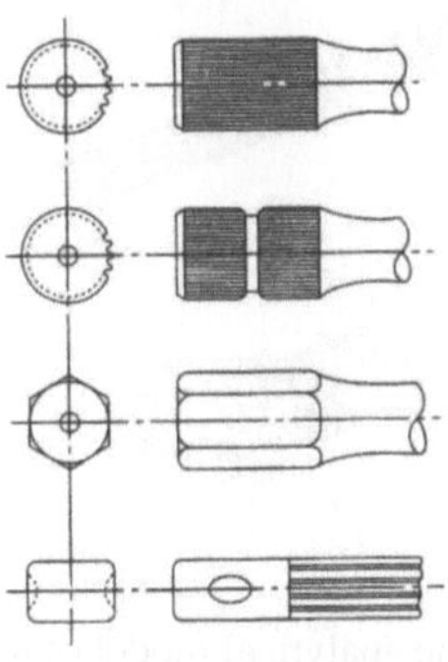

Figure 5.113: Examples of the end shape of torsion bars.

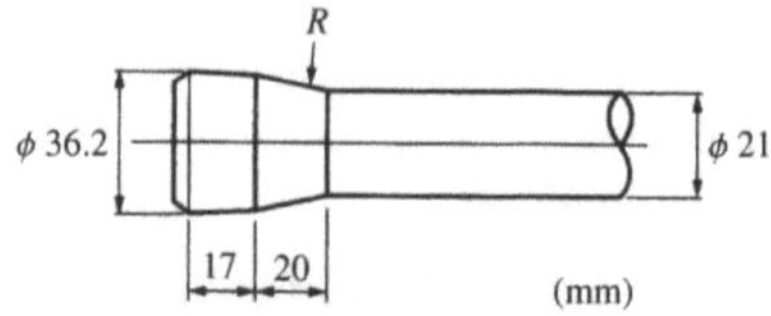


Figure 5.114: The end shape and dimension of end grip.

In torsion bar springs, the stress concentration takes place at end grips. The design that considers such stress concentration is indispensable, however, it has been carried out by a rule of thumb since the precise prediction of stress concentration can't be done using conventional design formulas.

The end of the torsion bar spring has been designed based largely on experience [34]. In the case of a serrated end, for instance, serration bottom diameter shall be 1.2 times or larger than the stem diameter.

FEM is an effective tool in verifying the validity of these past findings or of attempting a more reasonable design. An analytical example of stress concentration at the region *R* in Fig. 5.114 will be presented here briefly. *R* is assumed to be 10 mm and 30 mm.

**(1) Analytical model** Fig. 5.115 shows the analytical model.

In this model, 15-nodal solid elements shown in Fig. 5.103 are used. As can be seen from Fig. 5.115, the R region is divided into a finer mesh for the anticipated stress concentration. The analysis was carried out based on the condition that four nodes on the right tip surface were given a forced displacement equivalent to 30 degrees of torsional angle to the direction shown in the figure.

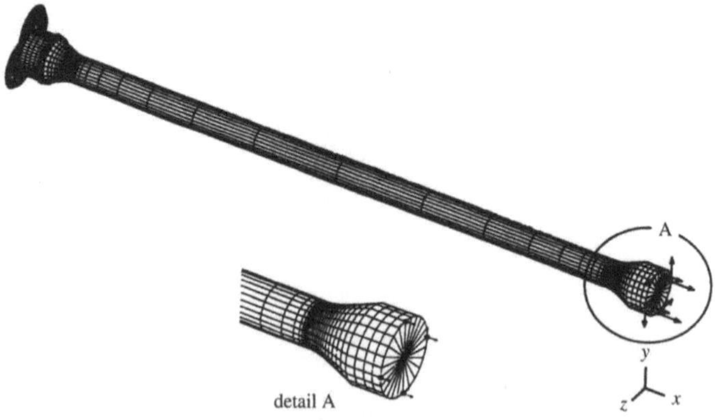


Figure 5.115: The analytical model of torsion bar spring.

**(2) Analytical result** The distribution of the maximum principal stress within the longitudinal cross section is shown in Fig. 5.116 and Fig. 5.117 for a 10 and 30 mm $R$ respectively.

In Fig. 5.118, the relation between the distance from tip end and maximum principal stress on surface is shown.

The maximum principal stress at the $R$ region shows approximately 5 % higher than the stress at stem region when $R$ is 10 mm, and approximately 2 % higher when $R$ is 30 mm. Since an evaluation of error is important, it is necessary to execute the analysis by finer discretization to find whether or not the solution converges. In other words, the discretization and the analysis must be repeated until the solution converges when adaptive mesh generation is not adopted into the FEM software used for this kind of analysis.

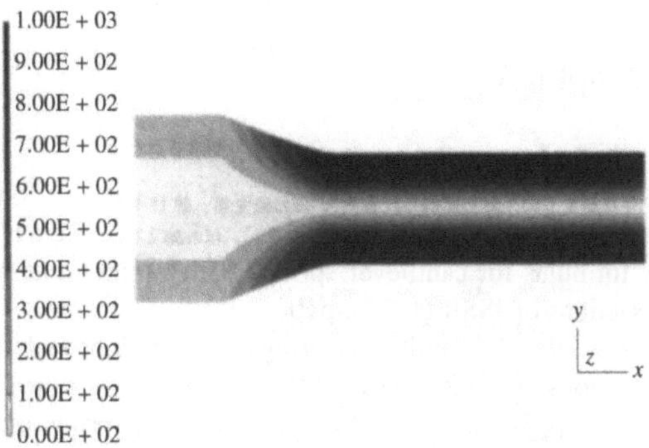


Figure 5.116: Distribution of maximum principal stress ($R$ = 10 mm).

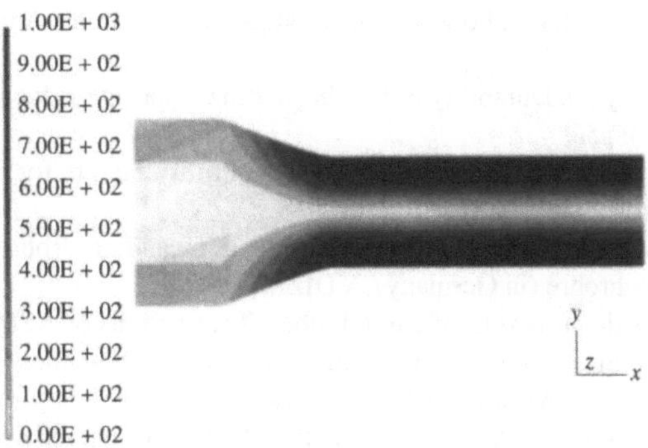


Figure 5.117: Distribution of maximum principal stress ($R$ = 30 mm).

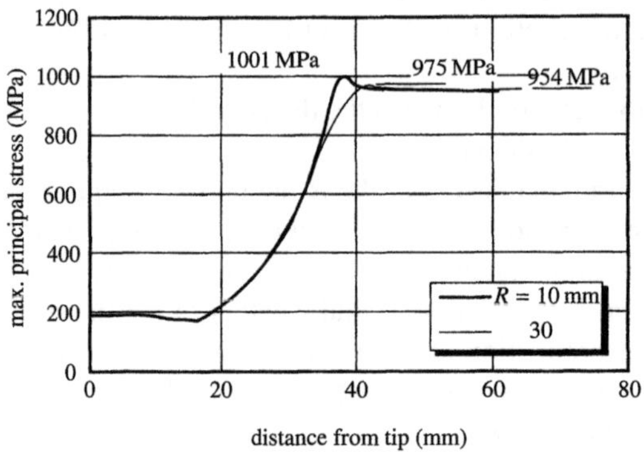


Figure 5.118: Distribution of maximum principal stress.

## Bibliography of Chapter 5


[1] JSSR, Bane 3rd ed. (in Japanese) 1982, Maruzen, p300

[2] Morita N et al, Displacement and stress analysis of wire spring composed of circular arc and straight line (in Japanese), Transactions of JSSR (1992), p47

[3] Oguti S, Utility formulas for cantilever springs with straight and curved portions (in Japanese), Transactions of JSSR (1972), p10

[4] Oguti S, Utility formulas for cantilever springs with straight and curved-portions (in Japanese), Transactions of JSSR (1976), p29

[5] Shimoseki M et al, Application of FEM to wire formed spring and stabilizer bar (in Japanese), Transactions of JSSR (1988), p41

[6] JSSR, Bane 3rd ed. (in Japanese) 1982, Maruzen, p320

[7] JSSR, Bane 3rd ed. (in Japanese) 1982, Maruzen, p264

[8] Inoue K, Design of the hollow stabilizer (in Japanese), Transactions of JSSR (1983), p46

[9] Higuchi J, Kamiya S, Durability of tubular stabilizer spring (in Japanese), Transactions of JSSR (1983), p55

[10] Dodge W G, Moor S E, Stress indices and flexibility factors for moment loading on elbows and curved pipe, WRC Bulletin 179

[11] Karman ThV (1911), Ueber die Formänderung dünwandiger Rohre, insbesondere federnder Ausgliechrohre (in Germany), VDIZ55, p1889

[12] Viegness I, Elastic properties of curved tubes, Transactions of ASME (1943), p105

[13] Sibuya M et al, Stress analysis of coiled spring with tapered ends (in Japanese), Technical Report of JSSR Meeting (1987 Fall), p45

[14] Shimoseki M, Fujinuma H, An isoparametric element for curved and twisted beams (in Japanese), Transactions of JSSR (1990), p25

[15] Shimoseki M, Hayasaka Y, Stress analysis of coil spring (in Japanese), Transactions of JSME Vo1. 48 435C (1982), p1774

[16] Shimoseki M, Sadakuni Y, Analysis of coil springs using joint element (in Japanese), Transactions of JSSR (1987), p1

[17] Hamano T, Tachikawa T, Analysis of compression coil springs by finite element method with joint elements (in Japanese), Transactions of JSSR (1993), p25

[18] Suzuki S, Muramatu T, Lateral deflection of coil springs used in automotive suspensions (in Japanese), Technical Report of JSSR Meeting (1994 Fall), p9
[19] JSSR, Bane 3rd ed. (in Japanese) 1982, Maruzen, p212
[20] Shimoseki M, Hayasaka Y, Stress analysis of coil spring (in Japanese), Transactions of JSSR (1984), p19
[21] Wahl A M, Mechanical springs (1963), McGAWHILL, p119
[22] JSSR, Design of springs 2nd ed. (in Japanese) 1978, Maruzen, p113
[23] Yoshimoto G, Hanabusa H, Torque calculation method of spiral spring (in Japanese), Transactions of JSME Vol. 18 (1952), p17
[24] Shimoseki M et al, On the numerical analysis of spiral springs (in Japanese), Transactions of JSSR (1988), p34
[25] JSSR, Bane 3rd ed. (in Japanese) 1982, Maruzen, p274
[26] Hamano T, Itou M, Analysis of leaf spring using finite element method with a joint element (in Japanese), Transactions of JSSR (1991), p39
[27] Watari A, On the deflection and the stress of a leaf spring (in Japanese), Transactions of JSSR (1952), p57
[28] Aoyama S, Yoshikawa K, Calculation of the stress distribution of conventional leaf springs for motor truck use (in Japanese), Transactions of JSSR (1977), p43
[29] Kodama M, Stress analysis of leaf spring (in Japanese), Transactions of JSSR (1955), p36
[30] NHK Spring Co. Ltd, Bane (in Japanese) 1995, Kogyo Chosakai Publishing Co. Ltd., p154
[31] Nishida M, Stress concentration (in Japanese) 1988, Morikita Publishing Co. Ltd., p309
[32] Reserchboard of JSME, Application of nonlinear Analysis by FEM (in Japanese), RC-74 Technicalreports I (1986), p139
[33] JSSR, Bane 3rd ed. (in Japanese) 1982, Maruzen, p283
[34] JSSR, Bane 3rd ed. (in Japanese) 1982, Maruzen, p267

# 6 EXPANSION OF ANALYTICAL HANDLING

Readers who have read through chapter 5 now have an idea of how effective FEM is to handle problems that are difficult or unsolvable with conventional formulas. In the previous chapter, we put emphasis on understanding fundamentals on FEM analysis. Consequently, we were narrowly restricted within the range of linear analysis and were unable to deal with many practical problems.

In this chapter, some of the problems which were omitted in the previous chapter but which are still considered important for the analysis of spring will be discussed. Almost all of these are nonlinear problems. A few nonlinear problems that were easy to handle and important for the analysis of the spring, were already discussed in chapter 5. Other types of complex issues will be discussed in this chapter.

To begin our discussion, an outline of the problem will be explained. Our first problem is the cross-section of a wire. This problem is technically linear but comparatively difficult. When a cross-section, other than a circular cross-section, is under torsional moment, we should consider the cross-sectional strain. To treat this phenomenon using the beam theory, we must take a strain function of the cross-section into account. Non-circular cross-section in coil spring and flattened hollow cross-section in stabilizer bar are typical cases.

Secondly, a contact problem including friction and large deformation will be discussed. These are non-linear problems that apply to all springs. However, the importance of the problem varies depending on spring types. In the case of a leaf spring, for instance, the large deformation and contact problems are important. In the case of a spring with nonlinear characteristics like the rolling bearing auxiliary spring, in particular, consideration of the contact problem is essential. The nonlinear movement of shackles cannot be treated without taking the large deformation problem into account. In addition, the hysteresis characteristic is accompanied by friction. To treat friction, the knowledge of relative displacement is necessary and the large deformation problem must be taken into account. In the case of the helical spring, on the other hand, contact problem is much more important than the large deformation problem. Contact between bearing seat and end coil or between coils takes place in helical springs. Consequently, the contact problem is essential in nonlinear helical springs. For disc and ring springs, linear analysis is applicable if friction is neglected. Nonlinear analysis is necessary when friction is to be taken into account.

Thirdly, elasto-plasticity as another nonlinear factor will be discussed. In general, it is in manufacturing process that the elasto-plasticity becomes an issue. The presetting, which is applied to almost all springs, is one of the major considerations.

An analytical treatment regarding the presetting has rarely been undertaken. When presetting is applied, a helical spring changes its height and a leaf spring changes its camber. Thus, the estimation of the change in coil height or camber caused by the presetting is an important issue in the design of springs. Presetting is also discussed in this chapter.

All the matters thus far are static problems, whether they are linear or nonlinear. Now, a spring is often used in a dynamic environment. A typical usage is an engine valve spring. In the previous chapter, the eigenvalue analysis of helical spring was discussed. As a final issue, the surging problem of valve spring will be discussed in this chapter considering contact.

## 6.1 STABILIZER BARS

The first problem with stabilizer bar is the complexity of its geometry.

As is described in chapter 5, it is impossible to prepare a formula for each aspect of complex geometry in advance. Fortunately, FEM, is very effective to handle a complex geometry. An objective model could be decomposed down into its fundamental shape, and the complexity of global geometry is not a bothersome question at all. This was discussed in chapter 2. This process was further described in chapter 5 and the reader might have understood that even a stabilizer bar with complex geometry can be treated comparatively easily in FEM analysis.

The second problem is the shape of the cross-section. There are two types in the stabilizer bar, solid and tubular. The tubular stabilizer bar is more susceptible to cross-sectional deformation than the solid stabilizer bar. Therefore, the tubular stabilizer bar has an inherent problem. One of such a problem is a flattened cross-section. The cross-section becomes oval shape by bending process, and spring characteristic together with stress level get affected accordingly. This problem belongs to linear analysis.

The other related issue is the collapse of cross-section. When the wall of tube is too thin, the cross-section might severely deform by force. This problem is in the range of nonlinear analysis. This issue will not be discussed any further in this chapter. The remaining issue of tubular stabilizer bar will be discussed in Paragraph 6.1.1.

The third problem is the evaluation of bush, which is common with both solid and tubular stabilizer bars. A bush is used to hold a stabilizer bar that is mounted on a car body. The bush material is mainly composed of rubber, but in some instances is comprised of metal. In the case of the metal bush, analysis can be executed based on general boundary condition. Examples of this were explained in chapter 5. In the case of rubber bush, such treatment is not possible. Rubber is elastic material and we need to take this characteristic into account. The treatment of such a bush will be discussed in Paragraph 6.1.2.

### 6.1.1 Tubular Stabilizer Bars

The flattening of tube affects load and stress characteristic of the tubular stabilizer bar[1]. Here, the treatment of tubular stabilizer bar by FEM will be discussed mainly on these issues.

The tubular stabilizer bar is composed of rectilinear section and curvilinear section (bent section). While the cross-sectional shape is circular in rectilinear section, it is somewhat flat at bent section due to bending in the forming process. The degree of flattening or the shape after flattening depends on bending condition. The prediction of the flattened shape requires elasto-plastic analysis during bending process. In this book, elasto-plastic analysis during bending process will be skipped. Only a flattened section will be studied. The problem of such a flattened section depends mainly on the selection of element and it does not matter whether the spring characteristic is linear or nonlinear. In the present analytical example, only an example of linear analysis will be presented.

In chapter 5, solid stabilizer bar was analyzed by beam element and tubular stabilizer bar by pipe element. The circular cross-section can be treated by these elements while the flattened cross section cannot. In addition to these elements, the shell or the solid element can be considered a possibility. In the case of the shell element, for example, the flattening of the cross-section will be handled properly as long as the wall thickness remains constant. In actual products, however, the thickness is uneven for the inside and the outside of the bending. The shell element is not appropriate to deal with these deviations of thickness. Based on such considerations, the solid element will be used for the modeling in this sample analysis[2]. An analytical example is shown below.

#### (1) Influence of a flattened cross-section

(a) Analytical model: Fig. 6.1 shows an analytical model of tubular stabilizer bar.

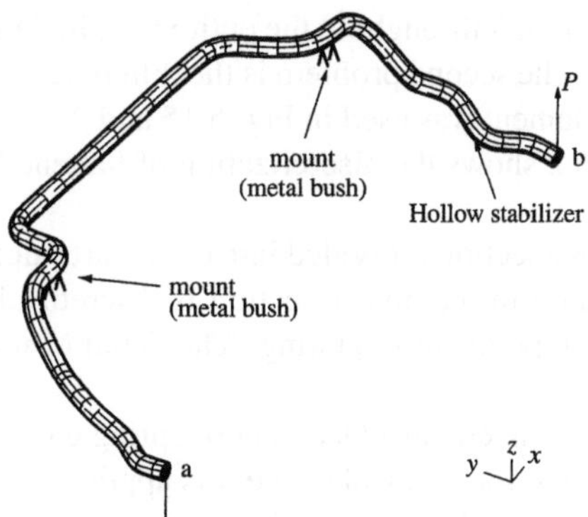


Figure 6.1: Analytical model of stabilizer bar mounted by a metal bush.

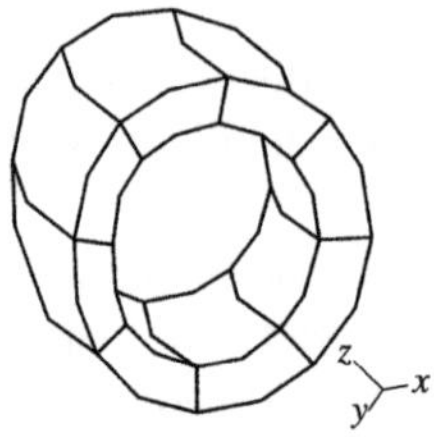


Figure 6.2: Discretization of flattened cross section at bent section.

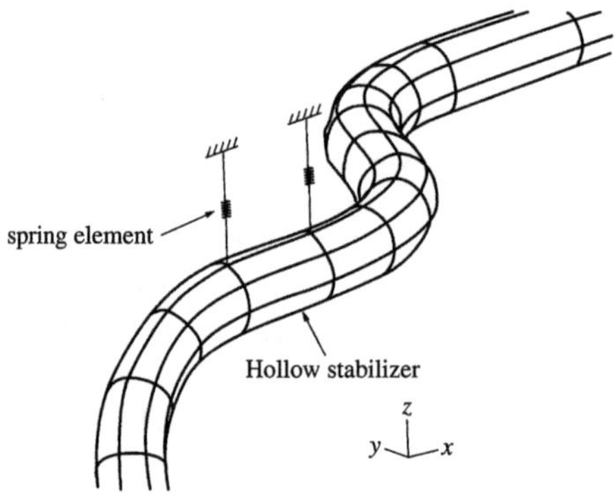


Figure 6.3: The detailed model of metal bush mount.

The analytical model is fundamentally the same as Fig. 5.15 in section 5.2. There are two differences, however, between Fig. 6.1 and Fig. 5.15. One is the difference in the shape of cross-section at bent section. While the shape of cross section in Fig. 5.15 is assumed to be circular throughout the entire length, flattened shape is located at bent area in Fig. 6.1. The second problem is the difference between elements used for the analysis. Pipe element was used in Fig. 5.15 and 20-nodal solid element was used in Fig. 6.1. Fig. 6.2 shows the discretization of flattened cross section at bent section.

In the figure, the cross-section is divided into eight parts along the circumference. The circular cross-section in rectilinear section is discretized in a similar way. A metal bush is used at the points of mounting. The detail of mount area is shown in Fig. 6.3.

For restraining condition, end displacement of spring element is set to zero in all directions. In addition, the forced displacement is applied to the nodes (point a and b, inverse to each other) in $z$ direction, as is shown in Fig. 6.1. This analysis was carried out on two types of stabilizer bars shown in Table 6.1.

Type 1 is a stabilizer bar with minimal bends. Type 2 has bends within the supporting span. The analysis was carried out on two occasions (flattened and non-

Table 6.1: Type of analytical stabilizer bars.

| | | Type 1 | Type 2 |
|---|---|---|---|
| tube dia. | | $\phi$ 22.21 mm | $\phi$ 22.2 mm |
| thickness | | 2.6 mm | 3.2 mm |
| flattening rate | First bent area | 9.46 % | 5.42 % |
| | Bent area 1 | – | 6.76 % |
| | Bent area 2 | – | 6.76 % |

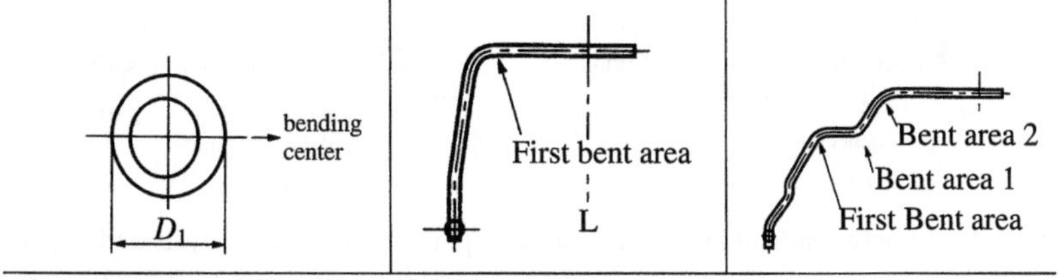


flattening rate $= 100 \times (D_0 - D_1)/D_0$
$D_0$: initial dia.
$D_1$: flattened dia.

flattened) for the sake of comparison. A linear analysis was executed.

(b) Analytical result: The deformed shape of type 1 and type 2 is shown in Fig. 6.4.

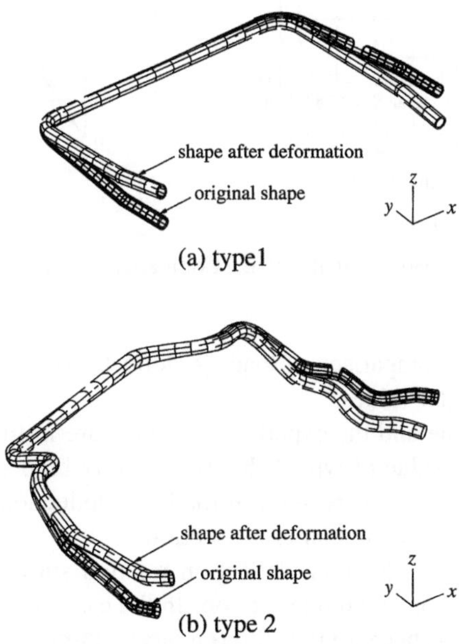


Figure 6.4: Deformed shape of tubular stabilizer bar.

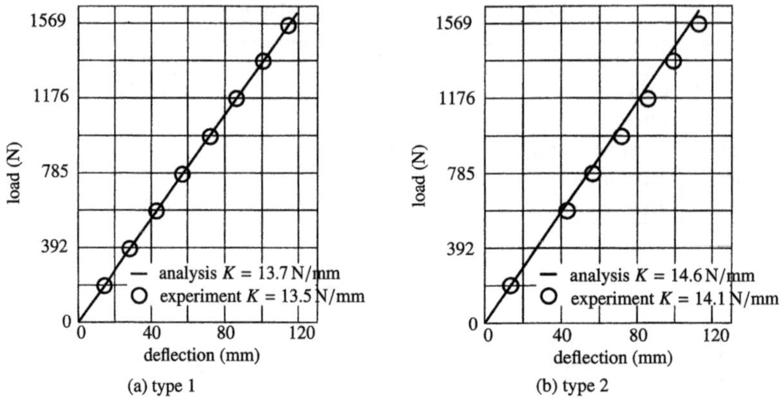


Figure 6.5: Relation between load and deflection (comparison between analysis and experiment).

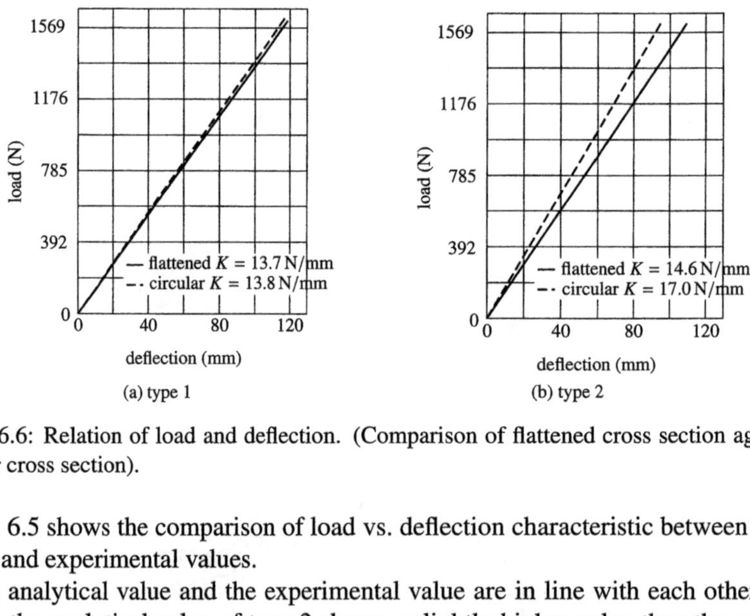


Figure 6.6: Relation of load and deflection. (Comparison of flattened cross section against circular cross section).

Fig. 6.5 shows the comparison of load vs. deflection characteristic between analytical and experimental values.

The analytical value and the experimental value are in line with each other, although the analytical value of type 2 shows a slightly higher value than the experimental value. In Fig. 6.6, the relation of load and deflection is compared between flattened cross-section and circular cross-section.

In the case of type 1, load vs. deflection relation shows almost no difference between flattened section and round section. In the case of type 2, however, the difference is large. This is because type 2 has a larger number of bends, and the effect of flattening reveals itself more significantly than in type 1. It can be understood

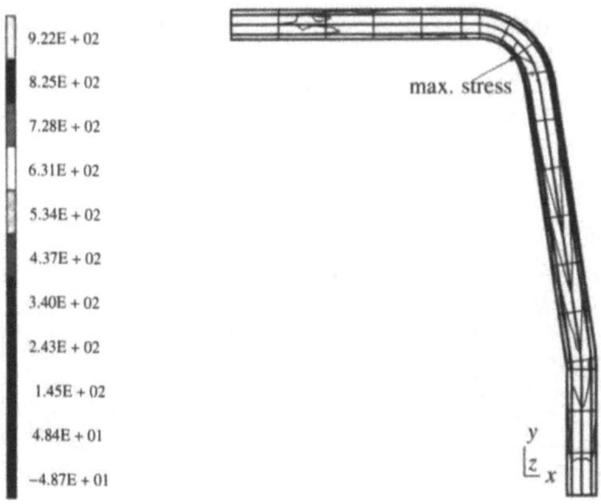


(a) type 1

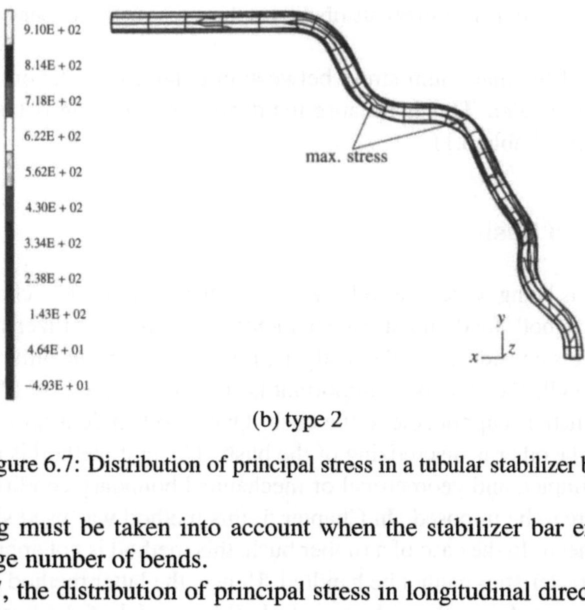


(b) type 2

Figure 6.7: Distribution of principal stress in a tubular stabilizer bar.

that flattening must be taken into account when the stabilizer bar encompasses a relatively large number of bends.

In Fig. 6.7, the distribution of principal stress in longitudinal direction is shown by contour lines.

In the case of the type 1, maximum stress is generated at shoulder. In the case of the type 2, however, maximum stress is generated not only at shoulder, but also at interior bend section. This is because the position of maximum stress is related to the whole geometry of the stabilizer bar.

Fig. 6.8 is a comparison of principal stress distribution at the bent area between circular and flattened cross-section.

In the case of type 1, the maximum stress of flattened cross-section is larger than that of circular cross section. The difference is clearly visible. In type 2, however,

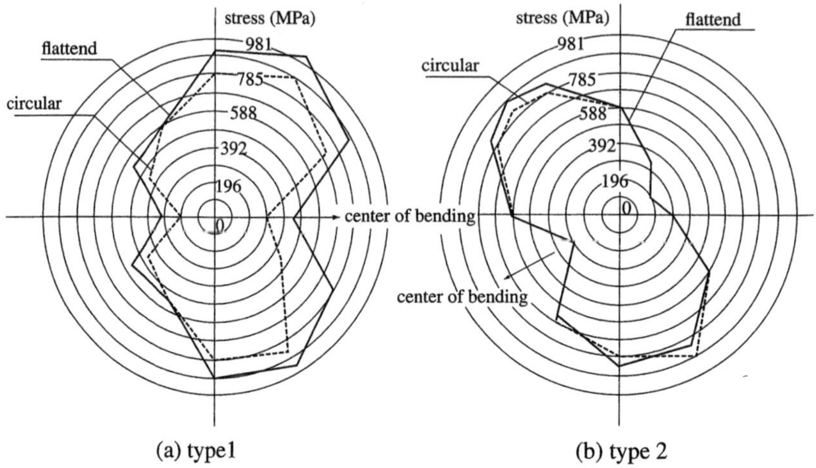


Figure 6.8: Principal stress distribution at the cross-section of bent area.

the difference of the maximum stress between circular cross-section and flattened cross-section is smaller. This is because the degree of flattening is larger in type 1 than in type 2 (see Table 6.1).

### 6.1.2 Effect of Bush

A rubber bush is being widely used for stabilizer bar support. The characteristic of this bush affects both load and stress characteristic of the stabilizer bar. Whether or not the bush is considered in the analysis, it will affect the accuracy of analysis. The way we handle the bush is an important issue to be decided. In FEM, there are two handling methods applicable to the bush. One is to handle it through boundary condition and the other is discretizing of the bush. The first method is more popular because it is simpler, and geometrical or mechanical boundary condition described in Chapter 1 are to be imposed. In Chapter 5, this method was used since the bush was made of metal. In the case of a rubber bush, this method is not appropriate since the elastic characteristics cannot be handled. Hence, the latter method is applicable. A simple procedure for such methods is to build a model of the bush using solid element with the physical characteristics of rubber.

However, the stabilizer bar must also be modeled by solid element if the bush is modeled in this way. The solid stabilizer bar cannot be modeled by simple beam element. So, the analytical model will be very complex. A substitute to avoid this complexity is to model the bush as a contact element. The contact problem can be considered, as is described in Chapter 2, as a problem of varying boundary condition. We can give elastic characteristic to contact element. So, the recommendable method is to model the rubber bush as a contact element that has a linear spring characteristic (linear spring element)[2], though it might be a crude approximation.

In this case, it is necessary to employ spring constant suited to the elastic characteristic. This technique is applicable to both solid element and beam element. The analytical example is described below.

**(1) Analytical model** An analytical model of tubular stabilizer bar gripped by rubber bush is shown in Fig. 6.9.

The stabilizer is modeled in the same manner as shown in Fig. 6.1 except that the grip is modeled by linear spring element. Fig. 6.10 shows a model of rubber bush.

The spring constant shall be assumed to be a linear spring element. Fig. 6.11 shows the installed condition of rubber bush and its spring characteristic.

As can be seen, the rubber bush has a dual characteristic depending on upward and downward directions. The spring constant of linear spring element in analysis is defined with this dual characteristic. All the remaining conditions in analysis follows subsection 6.1.1.

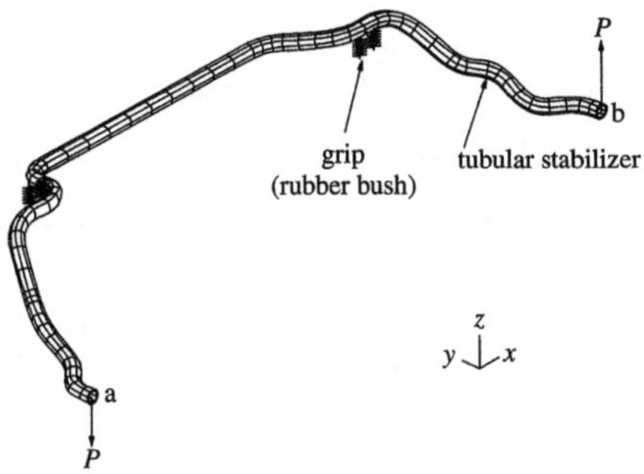


Figure 6.9: Analytical model of tubular stabilizer bar gripped by rubber bush.

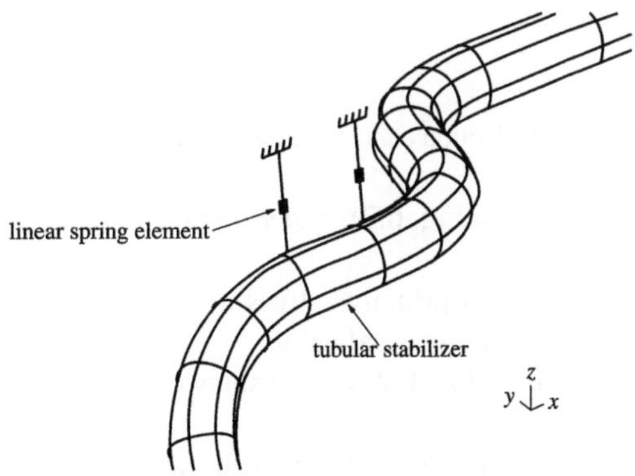


Figure 6.10: A model of rubber bush.

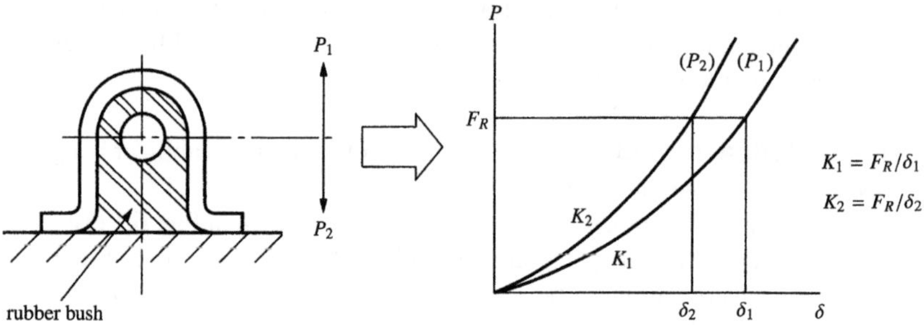


Figure 6.11: A rubber bush and its characteristics.

**(2) Analytical result** In Fig. 6.12, load-deflection chart is compared between analytical and experimental value for rubber bush case.

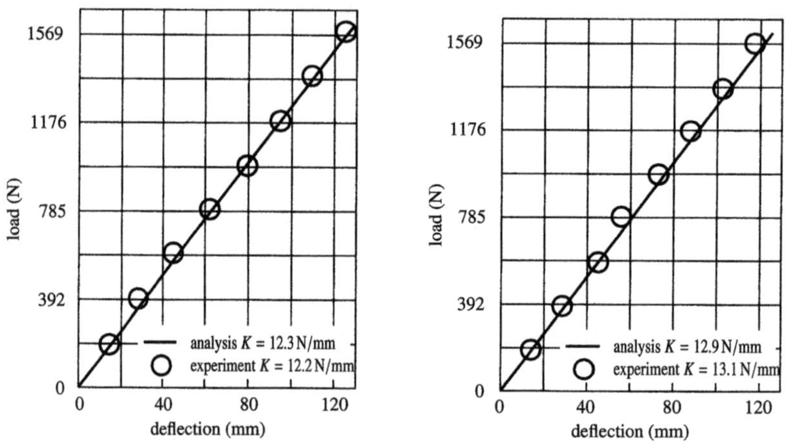


Figure 6.12: Lord-deflection chart (analytical versus experimental).

For both type 1 and 2 (refer to Fig. 6.7), experimental value and analytical value are in close agreement.

In Fig. 6.13, the distribution of principal stress in the cross section is compared between analytical value and experimental value.

The analytical value of stress distribution also shows harmony with experimental value.

In Fig. 6.14, load-deflection chart is compared between metal bush and rubber bush.

The spring constant of stabilizer bar with rubber bush has a tendency to show a lower value than that of metal bush for both type 1 and 2.

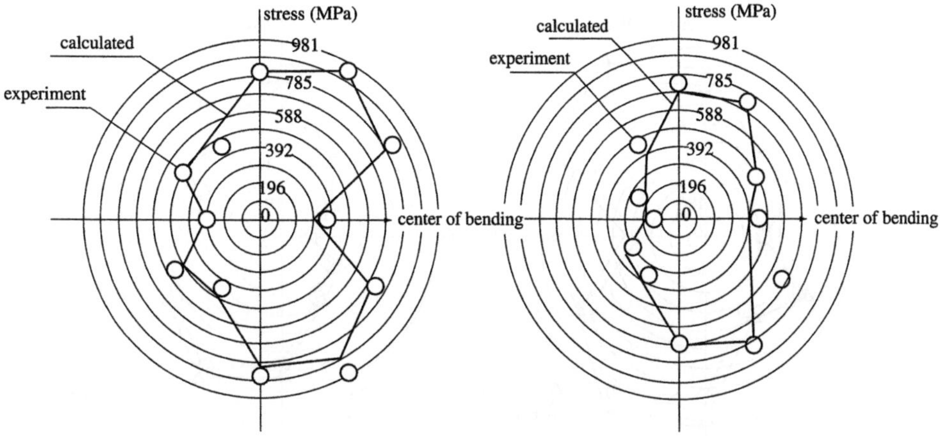


(a) type 1 (b) type 2

Figure 6.13: Principal stress on the cross-section of curved location.

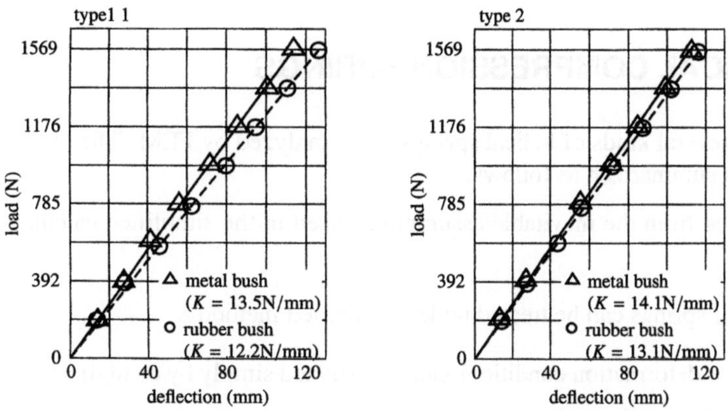


Figure 6.14: load-deflection chart. (metal bush versus rubber bush)

Fig. 6.15 shows the distribution of principal stress on the cross-section for two different bushes.

In both type 1 and 2, the stress of rubber bush has a tendency to show a lower value than that of metal bush. When rubber bush is used, the characteristic of rubber affects the characteristic of stabilizer bar. Therefore, it is necessary to assess the characteristic of bush for an accurate analysis.

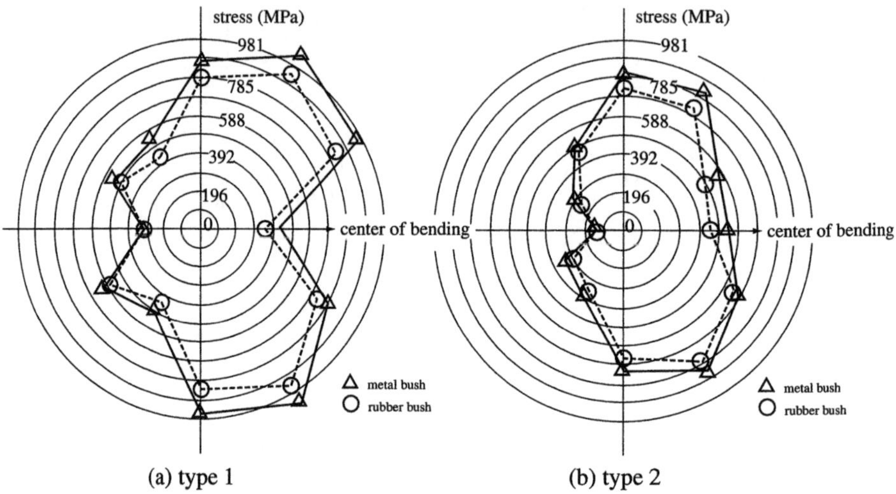


Figure 6.15: Distribution of principal stress on the cross-section of curved location. (metal bush versus rubber bush)

## 6.2 HELICAL COMPRESSION SPRINGS

In chapter 5, several kinds of helical springs were analyzed by FEM. The outcome of the study is summarized as follows.

- FEM is free from the inevitable assumptions used in the simplified calculation formulas.
- All the coil springs can be treated under the unified method.
- A variety of deformation conditions can be analyzed simply by changing boundary conditions.

The remaining problem concerning helical compression spring is the problem of contact between spring wires, as was noted in chapter 5. In helical compression spring, there occasionally is contact between retainer face and wire or between wires themselves. This contact problem is especially important for nonlinear springs like conical spring and unequal pitch spring. This is because nonlinear characteristic depends on contact between wires. The contact problem cannot be disregarded for springs with linear characteristics like the cylindrical spring. The load vs. deflection diagram of a cylindrical spring is somewhat nonlinear due to contact between wires and the change of pitch angle. In chapter 5, the position of contact was assumed to be fixed to treat contact problem within the scope of linear analysis. However, contact problem is essentially a nonlinear problem, and in actuality the contact position will change in accordance with load. It is necessary to take this contact position shift into account. In Paragraph 6.2.1, we will practice the change of contact position. The contact element introduced in Paragraph 4.1.5 will be used to this end.

In recent years, the popularity of miscellaneous cross-sections is increasing for lightweight springs. Rectangular wire or non-circular cross-sectional wire are good examples of these. A valve spring with non-circular cross-section is a new type of spring that has been attracting more attention in recent years. In chapter 5, the circular cross-section was treated by beam element. Rectangular or non-circular cross-section, however, cannot be treated with simple beam element. In Paragraph 6.2.2, a non-circular cross-section will be treated by solid element.

Presetting is applied when the anti-sag property of a spring is required. The presetting has been regarded as a problem of manufacturing. Therefore, an analytical report of presetting is rare. The presetting is an elasto-plastic problem and cannot be treated within the range of linear analysis. This presetting will also be treated by beam element in Paragraph 6.2.3. Lastly, engine valve springs receive a dynamic load through driving cam. Surging always becomes a problem in engine valve spring design. Dynamic analysis is necessary to treat surging. The analysis of the eigenvalue treated in Chapter 5 was aimed at nondamping free vibration among dynamic analyses. Since surging is a damped forced vibration, the problem of surging will be further discussed in Paragraph 6.2.4.

### 6.2.1 Problem of Contact

There are two sort of contact in considering helical compression springs. One is the contact between retainer face and wire and the other is the contact between wires. The former is a problem concerning all helical compression springs, regardless of whether the spring characteristic is linear or nonlinear. The latter is a typical problem with nonlinear springs[3]. When we try to solve the contact problem by expanding conventional simplified formula for helical springs, we have no choice but to follow the provided assumptions[4, 5, 6, 7, 8, 9] as follows:

1) For the contact between retainer face and wire, deformation is analyzed only on active coils, neglecting the end turn.

2) Regarding contact between wires, the contact starts from the end of active coils and proceeds further on. Once the contact takes place, there is no deformation on that area.

These preconditions are not always applicable to the actual behavior of helical springs. FEM is an adequate method to handle contact problem and does not depend on preconditions mentioned above.

In FEM analysis, all the contact condition can be treated in the same manner[10, 11]. The deformation of end turn can be evaluated too. An analytical example will be explained below.

**(1) Analytical model** Beam element and solid element will be presented here.

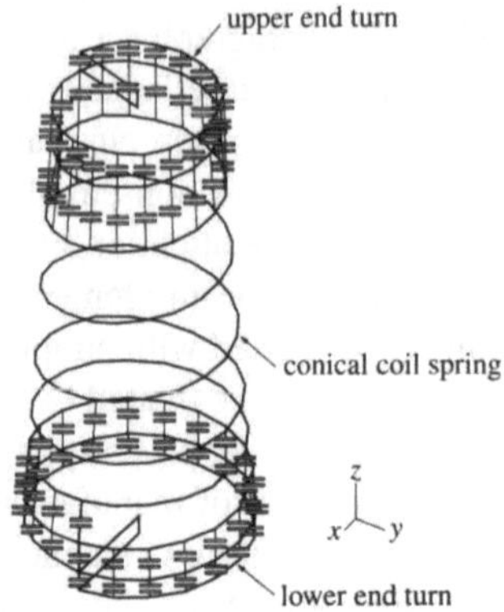


Figure 6.16: An analytical model for a conical spring. (beam element)

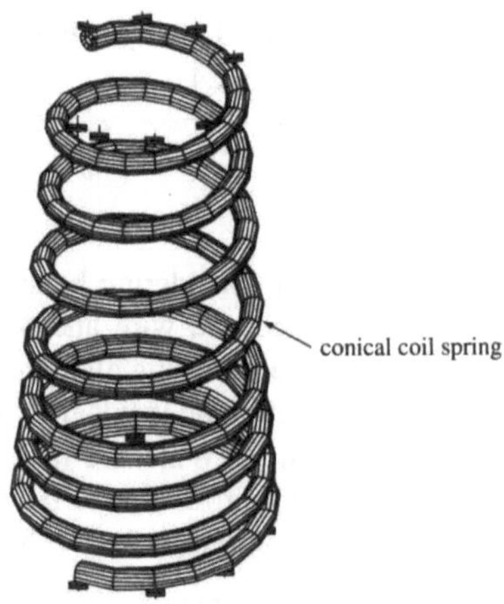


Figure 6.17: An analytical model for a conical spring. (solid element)

Fig. 6.16 shows the analytical model when a conical spring is modeled by two-node three-dimensional beam elements.

In chapter 5, only contact between retainer face and wire was considered. In the present analytical model, however, the contact was considered not only between retainer face and wire but also between wires themselves using contact elements.

In Fig. 6.17, the same conical spring as shown in Fig. 6.16 is modeled by 20-nodal solid elements.

The number of divisions per coil is 20 for each calculation model. The number of divisions on cross section is 8 in circumferential direction. When the beam element is used for modeling, the output is limited to forces in the structural member, and it is necessary to prepare an equation for stress. When solid element is used for

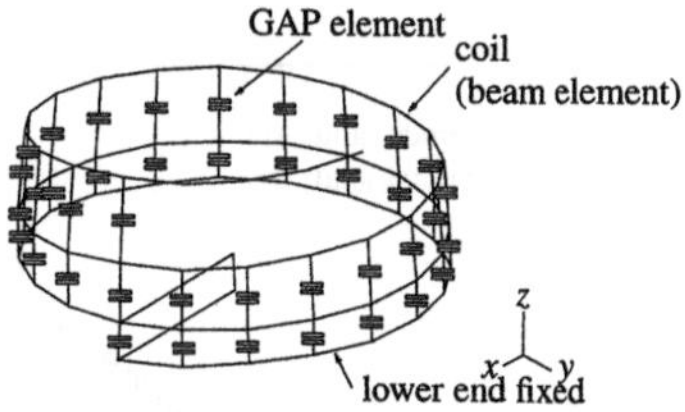


(a) Analysis model by beam element

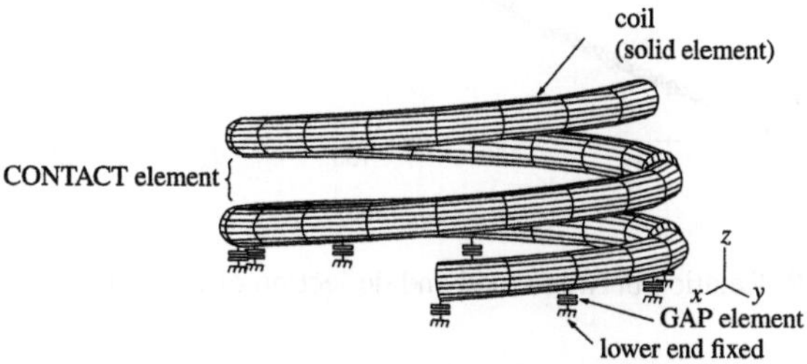


(b) Analysis model by solid element

Figure 6.18: Detailed model of contact area.

modeling, it is not necessary to prepare an equation for stress analysis since the stress analysis is a part of the process of constructing element stiffness matrix. As far as contact elements are concerned, constant-direction GAP elements are used for beam element model.

In the case of solid element model, constant-direction GAP elements are used between retainer face and wire too. However, a joining element (named CONTACT in the software MARC) is used between wires themselves. Fig. 6.18 shows the detail of contact area where GAP elements are inserted in a beam and solid element model.

The constant-direction GAP element can be used not only between beam elements but also between solid elements. This is an advantage of using this element in handling a contact problem. For the boundary condition, lower retainer is fixed and upper retainer is given a forced displacement. Nonlinear analysis is executed with friction at the contact area being neglected.

**(2) Analytical result** Fig. 6.19 shows the relation between load and deflection.

The analytical value is in good conformity with the experimental value in both cases. The transient section from the first spring constant to the second spring constant, where the nonlinearity is observed, also shows good correspondence with each other.

The relation between load and deflection of a cylindrical spring is shown in Fig. 6.20.

In Fig. 6.20, we can be observe that even a cylindrical spring, which is considered to have linear characteristic, shows a nonlinear characteristic at the initial stage of

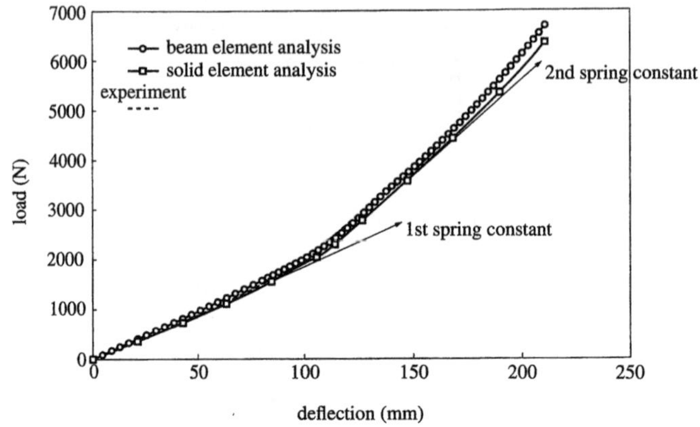


Figure 6.19: Relation between load and deflection of conical spring.

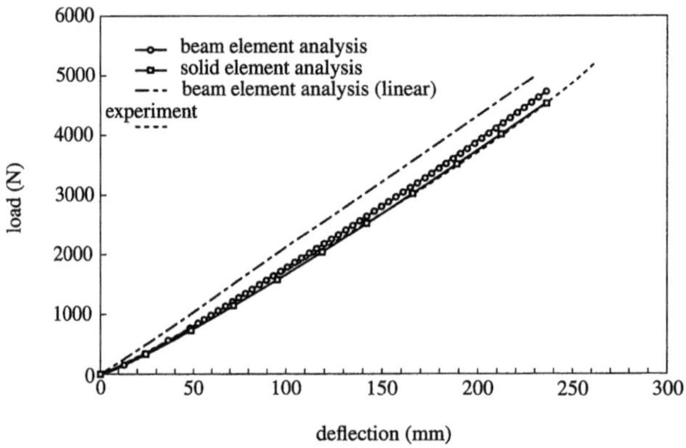


Figure 6.20: Relation between load and deflection of cylindrical spring.

loading. This is caused by contact between retainer face and end coil. The spring constant determined by linear analysis is higher than nonlinear analysis since only the active coils are used in the linear analysis, neglecting the contribution of end turns. Obviously, the linear analysis of active coils has a limited accuracy.

Fig. 6.21 shows the deformed state of a helical spring when beam elements are used.

Solid circles denote the location of contact. As to the deformation, the analytical value is in good agreement with the experimental value. Fig. 6.22 shows the contact state with regard to the load condition.

The abscissa shows the position of the node expressed by the coil number and solid circles show node points in contact. Observing the contact nodes for each load in Fig. 6.22, it can be said that contact nodes are not continuously lined. The

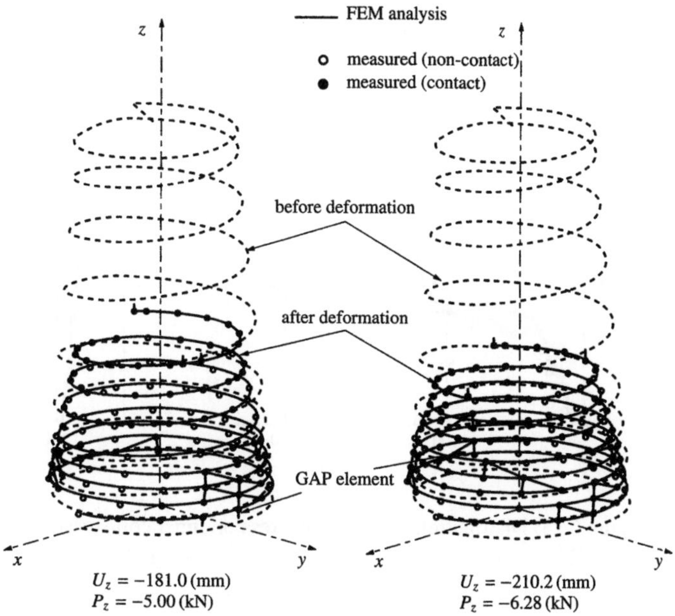


Figure 6.21: Deformation and contact state.

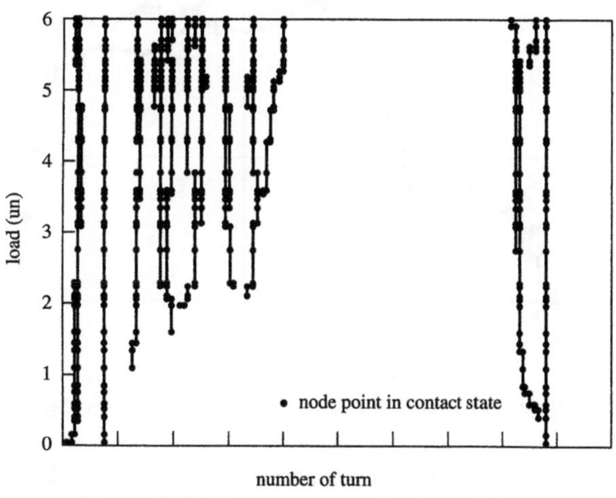


Figure 6.22: Contact state vs. load condition.

conventional precondition "*The contact is assumed to progress continuously and successively from the end*" is not valid as can be seen from this figure.

Fig. 6.23 and Fig. 6.24 show the stress distribution of the solid element analysis

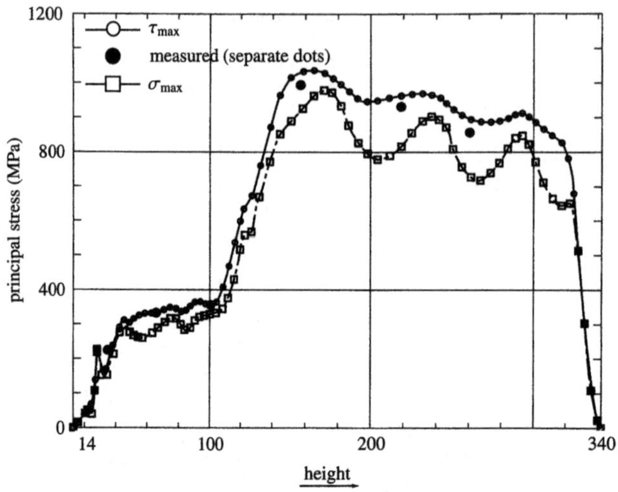


Figure 6.23: Distribution of principal stress on inner coil surface.

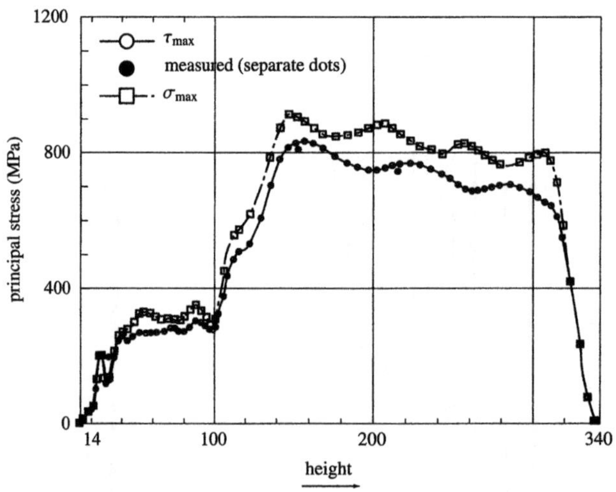


Figure 6.24: Distribution of principal stress on outer coil surface.

on inner and outer coil surface. The comparison of analytical value and experimental value can be seen too.

The appearance of a wavy distribution is similar to the analytical result of the beam element analysis in chapter 5. The abscissa is the height ($z$ coordinate value) of each coils. The reason why the stress is low within 100 mm is that the contact takes place in this area.

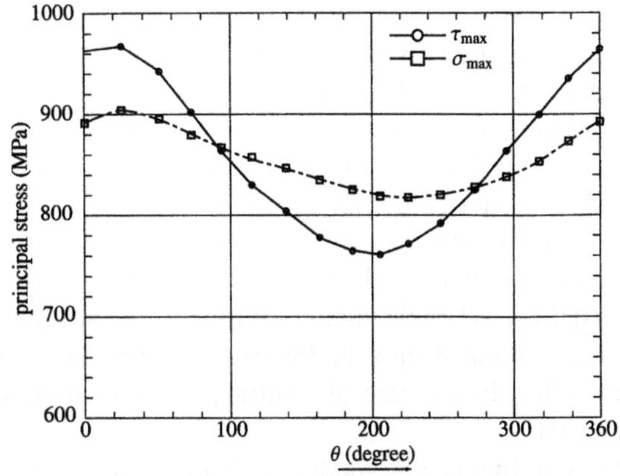


Figure 6.25: Distribution of principal stress on the circumferential surface of wire. (non-contact area)

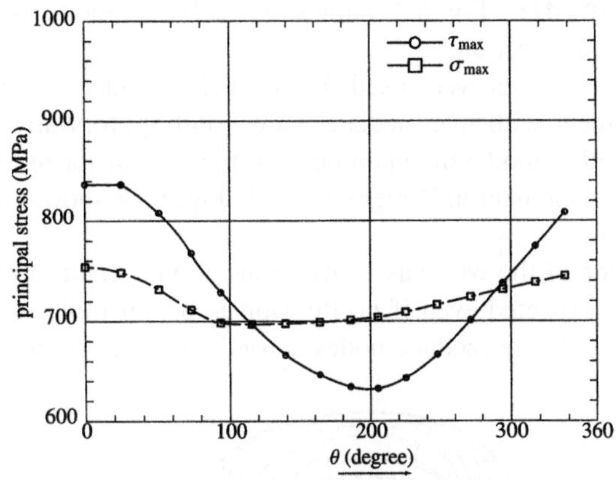


Figure 6.26: Distribution of principal stress on the circumferential surface of wire. (contact area)

Fig. 6.25 and Fig. 6.26 show the stress distribution on the circumferential surface of wire.

Due to the effect of curvature, the maximum shearing stress on the inside diameter of the coil ($\theta = 0$ degrees) is higher than the outside diameter of the coil ($\theta = 180$ degrees). In general, the peak value of the maximum shearing stress on the circular cross-sectional wire takes place on the inside diameter of the coil.

### 6.2.2 Non-circular Cross Section

There are many shapes of wire cross-section for helical compression springs other than a circle. A rectangle for example, is commonly used, and many arbitrary shapes are utilized[12, 13]. When they are used in a helical compression spring, the cross-section undergoes torsion. The problem in treating the arbitrary cross-section is that the cross-section does not hold flatness when it is twisted. To formulate this phenomenon based on beam theory, a distortion function must be defined. In the case of arbitrary cross-section[14, 15] of curved beam, this distortion function is very complicated to define. When the distortion function of straight beam is solved numerically, and is incorporated into the theory of curved beam, then the analysis becomes possible[16]. In the case of arbitrary cross-section, solid element is superior to beam element.

Any shape of cross section can be modeled with exactness using solid element. In this case, the distortion of the cross-section will be naturally considered without any special consideration during the process of the analysis. It can be regarded as a unified method. An analytical example will be presented here.

**(1) Analytical model** Fig. 6.27 shows the analytical model of a helical spring with an elliptical cross-section.

20-nodal solid elements were used for the FEM model. This element is an isoparametric element with intermediate nodes, and appropriate for modeling of curves. The method to model the main body of the helical spring is the same as in the case of the solid element in Paragraph 6.1.1. Fig. 6.28 shows the discretization of the cross-section.

The cross-section of the wire has 16 divisions along circumference. A satisfactory result can be expected even if no division is put in radial direction because 20-nodal element with intermediate nodes is used. If the eight-nodal solid element

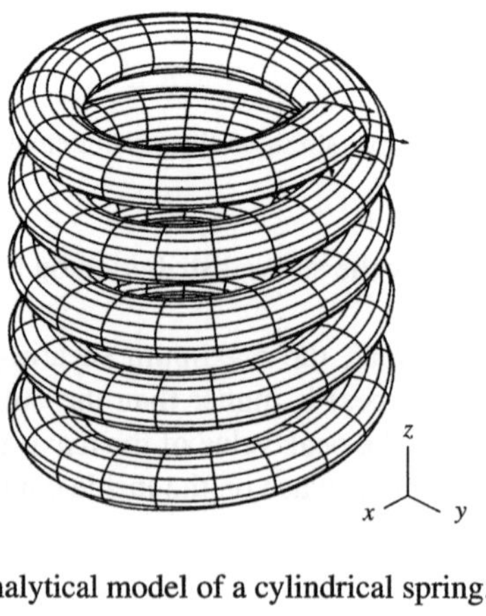


Figure 6.27: Analytical model of a cylindrical spring. (active coils)

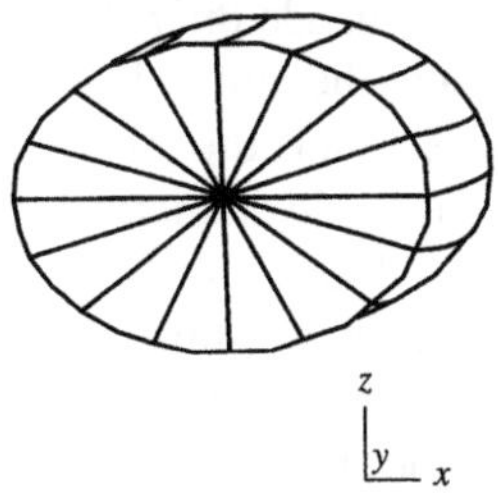


Figure 6.28: Discretization of cross-section.

is used in modeling, the division of two or more in radial direction is necessary. Contact was taken into account in Paragraph 6.1.1, but it is not the case in this paragraph. Therefore, the analysis was carried out by linear analysis taking only the active coils.

As for boundary condition, lower end turn was restrained and upper end turn was given a forced displacement. The analysis was carried out on four cases of cross-sectional aspect ratio ($a/b$). The aspect ratio "$a/b = 1$" corresponds to a circular cross-section. The weight of the spring remains constant since cross-sectional areas are chosen to be the same for all aspect ratio cases.

**(2) Analytical result** Fig. 6.29 shows the load and deflection chart.

It shows a linear characteristic because of the linear analysis. The spring constant shows a tendency to increase as the cross-section nears a circular shape ($a/b = 1$).

Fig. 6.30 and Fig. 6.31 show the stress distribution when aspect ratio of the wire

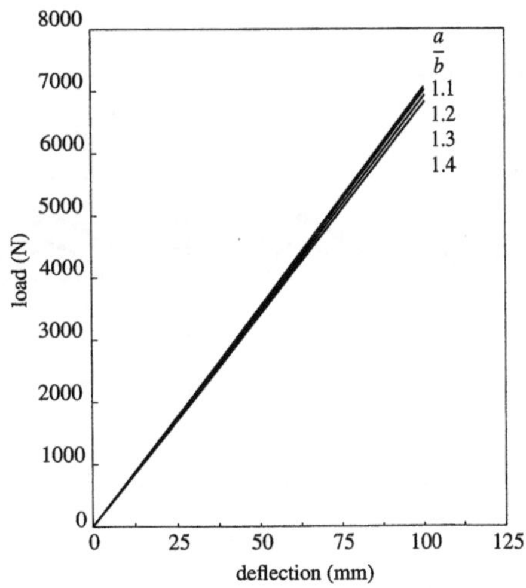


Figure 6.29: Load and deflection chart for a helical spring with an elliptical cross-section.

$(a/b)$ is 1.1, i.e., the cross-section is nearest to a circle.

Fig. 6.30 shows the stress at the inside diameter of coil and Fig. 6.31 at the outside. Fig. 6.32 and Fig. 6.33 show the stress distribution when aspect ratio of the wire $(a/b)$ is 1.4, i.e., the cross-section is the most flattened.

Fig. 6.32 shows the stress at the inside diameter of coil and Fig. 6.33 at the outside diameter. In both cases of $a/b = 1.1$ and 1.4, their stress distribution shows a wavy distribution by the influence of the pitch. This is similar to the stress distribution in Paragraph 6.1.2. When the two springs with the aspect ratio $a/b = 1.1$ and 1.4 are compared, the absolute value of the stress of the latter is lower than the former as a

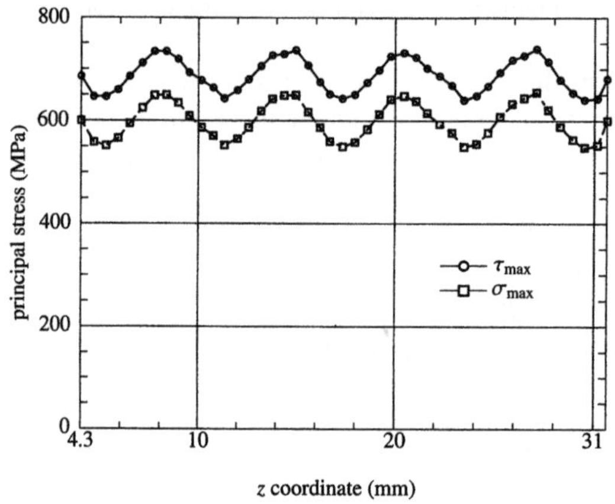


Figure 6.30: Distribution of principal stress at the inside diameter of coil. (Aspect ratio $a/b = 1.1$)

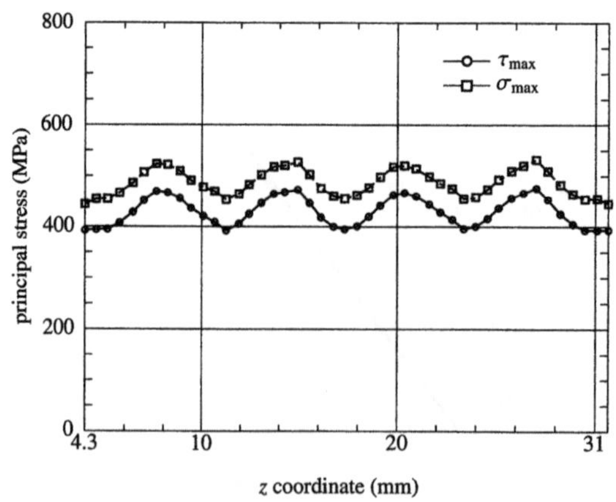


Figure 6.31: Distribution of principal stress at the outside diameter of coil. (Aspect ratio $a/b = 1.1$)

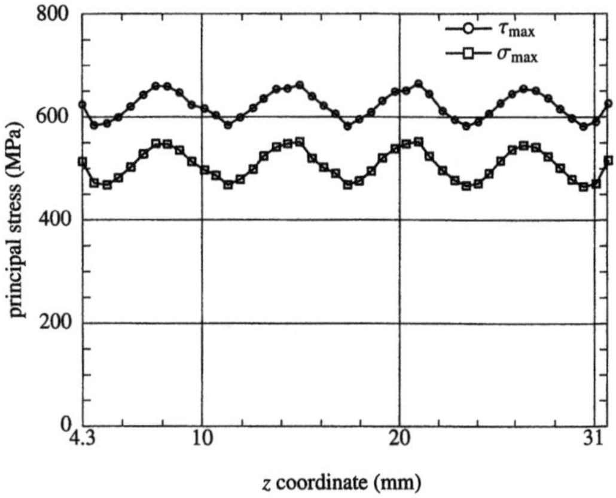


Figure 6.32: Distribution of principal stress at inside diameter of coil. (Aspect ratio $a/b$ = 1.4)

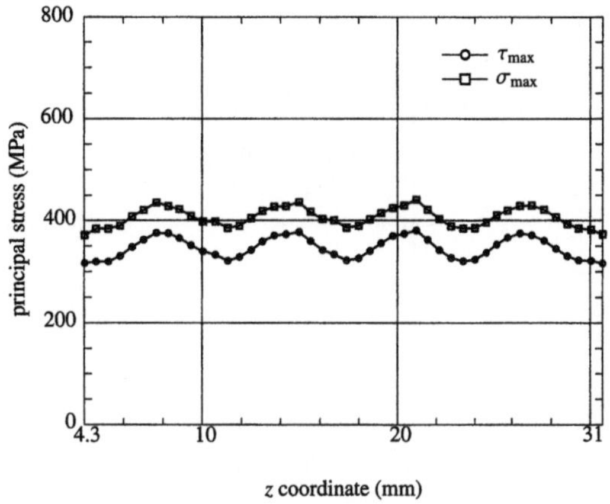


Figure 6.33: Distribution of principal stress at the outside diameter of coil. (Aspect ration $a/b$ = 1.4)

whole. The reason is that the spring constant for $a/b$ = 1.4 is lower than $a/b$ = 1.1, as is described above, and hence the reaction force received by the former spring is smaller than that of the latter when the same amount of forced displacement is applied as a boundary condition.

Figures from Fig. 6.34 to Fig. 6.37 show circumferential stress distribution in the case of both $a/b$ = 1.1 and 1.4.

Fig. 6.34 and Fig. 6.36 show the distribution of maximum shearing stress on the cross-section indicated by contour line. Fig. 6.35 and Fig. 6.37 show circumferential stress distribution on the surface. The location of the cross section was chosen at

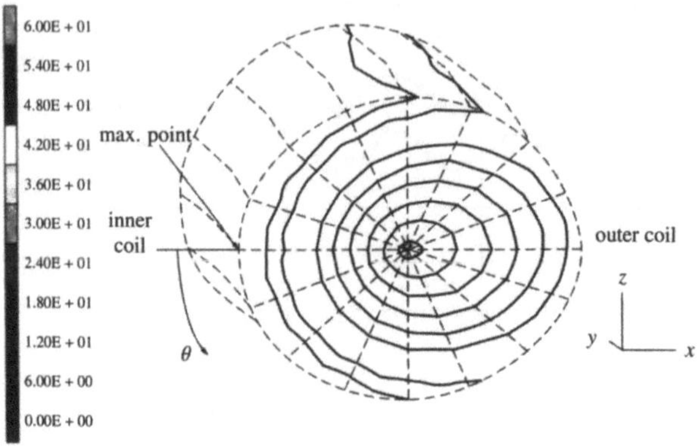


Figure 6.34: Distribution of the maximum shearing stress on the cross-section. (Aspect ratio $a/b = 1.1$)

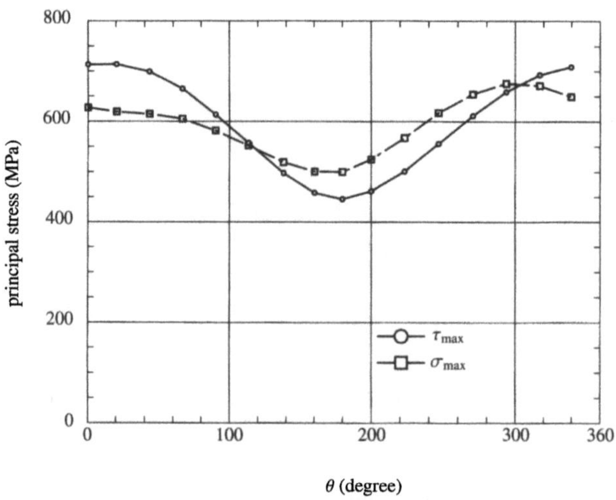


Figure 6.35: Distribution of surface principal stress. (Aspect ratio $a/b = 1.1$)

which the stress seems to take the highest value in the distribution of the principal stress in Fig. 6.30 and Fig. 6.32. When Fig. 6.34 and Fig. 6.35 are compared, it can be observed that the circumferential position of max. stress is different. In the case of near-circular cross-section ($a/b = 1.1$), the position of the maximum value is near the inside or the outside diameter of the coil. In the case of collapsed-elliptic cross-section ($a/b = 1.4$), the position of the maximum stress is not at the polar, but at the position slightly shifted from that position. This implies that it is possible to change the stress distribution within the cross-section and to carry out the optimum stress design by changing the cross-section shape of wire. From such

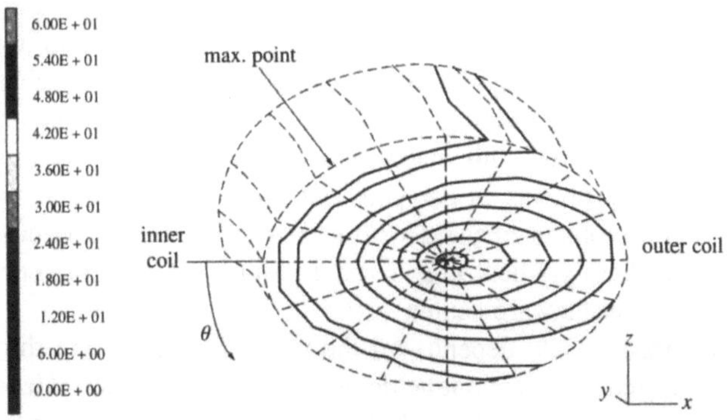


Figure 6.36: Distribution of the maximum shearing stress on the cross-section. (Aspect ratio $a/b = 1.4$)

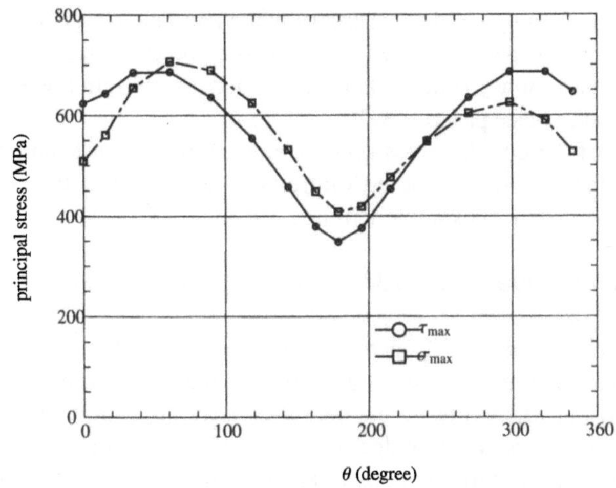


Figure 6.37: Distribution of surface principal stress. (Aspect ratio $a/b = 1.4$)

a viewpoint, non-circular cross-sectional wire, including the elliptic cross-section shapes are obtaining more attention from design engineers.

### 6.2.3 Presetting

When the load is applied to a spring, the spring deforms elastically until the surface of wire reaches yield point and begins to deform plastically. Presetting is an operation to give the spring this plastic deformation beforehand. Since presetting is thought to be a manufacturing process, it has not been considered to be an important problem in design. As an exception, the analysis of non-linear characteristic springs can be referred to[17]. However, presetting changes the height or coil diameter of the spring and affects the shape of the product. Therefore, it is an important design problem to analyze presetting and to foresee the amount of these changes.

Now, the problem of presetting is to analyze an elasto-plastic behavior as a material nonlinearity. By FEM, this elasto-plastic problem could be solved with ease. Among important problems in carrying out the elasto-plastic analysis of helical spring, there is a problem of yield within cross-section. In the case of compression spring, the stress gradient is generated radially in it's cross-section and yield proceeds from surface toward center.

The evaluation of yield depends on the type of element. In the case of beam element, the discretization cannot be done for the cross-section. Hence, the yield is evaluated by the use of either the yield function concerning moment[18] or stress at the predetermined integral point within the cross-section limiting the applicability to a narrower area.

In solid element, on the other hand, the cross-section of wire can be discretized and, yield within the cross-section is evaluated for each element. In short, no special consideration concerning yield is necessary when solid element is used. Presetting can be treated in this manner. When the yield is to be evaluated for each element and the analytical accuracy is important, subdivided discretization will be necessary because yield takes place progressively.

In the case of the circular cross-section, presetting can be handled by a beam element with ease. However, the curvature and the torsion of the coil cannot be considered. The analytical example will be presented here.

**(1) Analytical model** Fig. 6.38 shows an analytical model of a cylindrical spring.

Two-node three-dimensional beam elements were used for the modeling. The modeling of the main body of the spring is fundamentally similar to the case in subsection 6.1.1 in which a beam element was used. The important point for the elasto-plastic analysis is how to judge the yield. When a beam element is used, the judgment of the yield is carried out by stress at integral points. Fig. 6.39 shows the integral points for a two-node three-dimensional beam element used in the present analysis.

In this element, curvature and torsion are not considered.

With regard to contact, the retainer face was modeled by a rigid surface and the GAP element was inserted between the retainer and wire, and also between wires.

As to the restraining condition, the rigid surface of the lower side was fixed and the upper side was given a forced displacement, while they were kept parallel during the displacement. The analysis was carried out on examples No.1 and No.2 with different specifications. To carry out the elasto-plastic analysis, the diagram of true stress vs. logarithmic-strain diagram is necessary as an input data to judge the yield. There are two kinds of diagrams for this purpose. One is the diagram of experimental data by the tensile test. The other is an approximated diagram. These diagrams

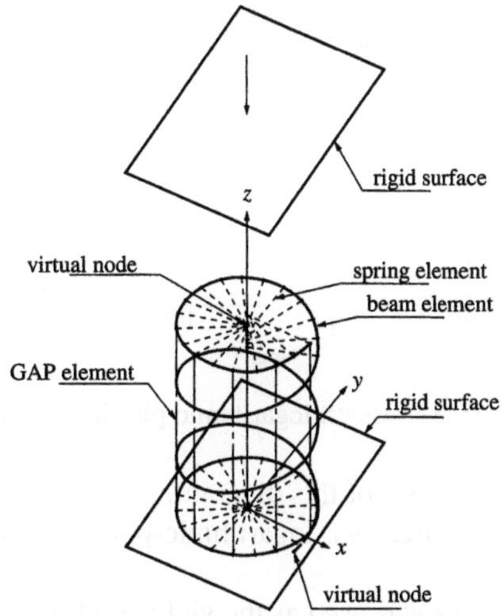


Figure 6.38: An analytical model of a cylindrical spring.

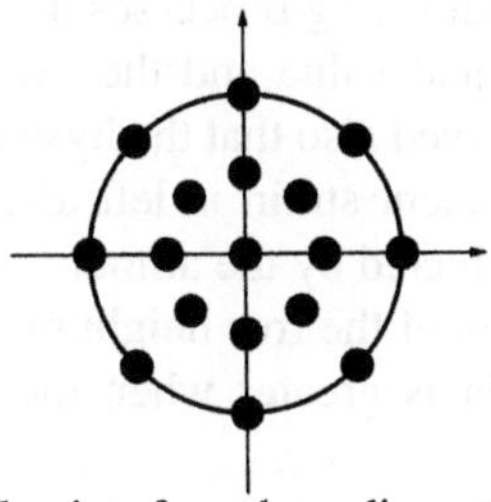

Figure 6.39: Integral points for a three-dimensional beam element.

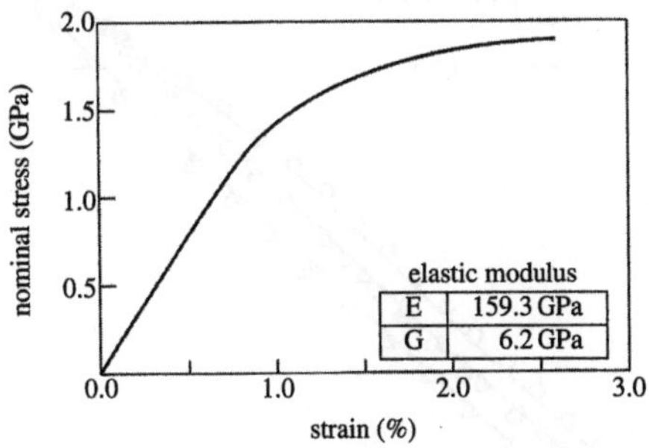


Figure 6.40: Nominal stress vs. nominal strain diagram. (experimental value)

vary depending on the material giving a big impact to the analytical result. It is recommended to use the experimental diagram if possible. An experimental diagram will be used in the present analysis. Fig. 6.40 shows the nominal stress vs. nominal

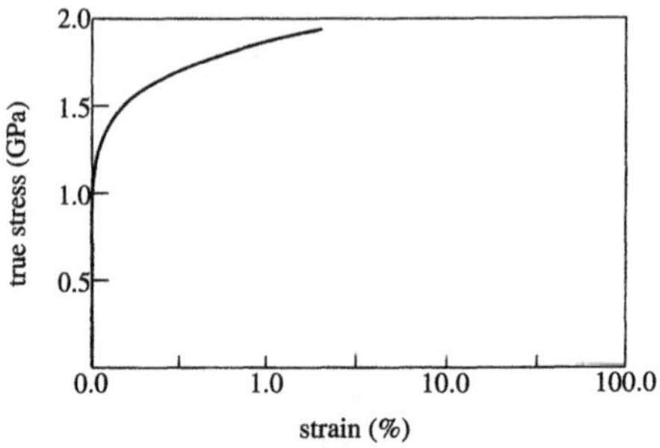


Figure 6.41: True stress vs. logarithmic-plastic strain diagram.

strain diagram, which is the basis of this diagram.

Fig. 6.41 shows the true stress vs. logarithmic-plastic strain diagram converted from Fig. 6.40.

Mises's yield stress theory was used as the yield condition.

**(2) Analytical result** Fig. 6.42 and Fig. 6.43 show the load vs. deflection characteristic during loading and unloading processes in the presetting.

In each process, the analytical value and the experimental value show a good correspondence. It can be observed also that the hysteresis takes place due to elasto-plastic deformation and permanent strain is left after unloading. The height and diameter after presetting are affected by the amount of this permanent strain.

Table 6.2 shows the variation of the free height by presetting.

The variation of free height is greater when the amount of presetting load is

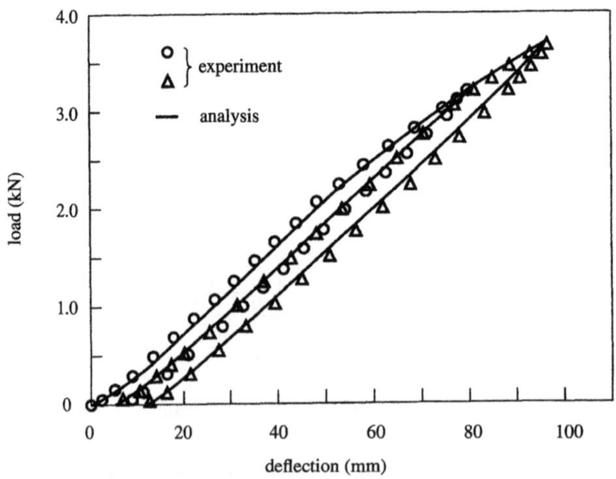


Figure 6.42: Relation between load and deflection (No.1).

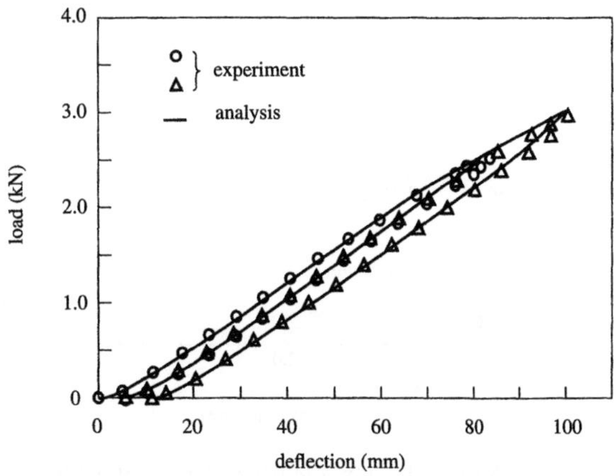


Figure 6.43: Relation between load and deflection (No.2).

Table 6.2: Variation of free height by presetting.

| No. | deflection (mm) | (mm) experiment | analysis |
|---|---|---|---|
| 1 | 79.0 | 6.3 | 5.1 |
| | 96.3 | 12.3 | 12.0 |
| 2 | 83.9 | 5.0 | 4.3 |
| | 100.0 | 10.7 | 10.3 |

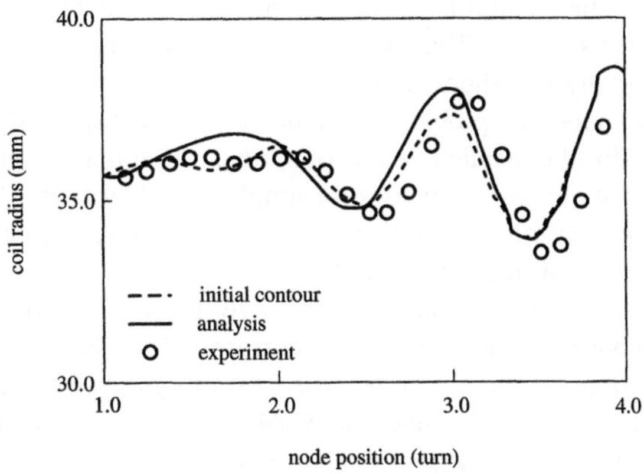


Figure 6.44: Variation of coil diameter by presetting (No.1).

increased. Fig. 6.44 and Fig. 6.45 show the variation of coil diameter, so-called bowing, by presetting.

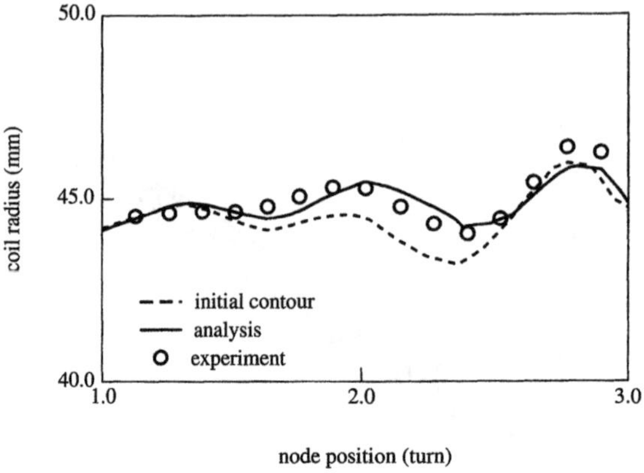


Figure 6.45: Variation of coil diameter by presetting (No.2).

The distribution of analytical value is in good agreement with that of experimental value. The amount of coil diameter variation by presetting fluctuates along with coil position. This phenomenon is caused by an uneven plastic deformation on the surface of wire. It could be emphasized that the presetting is an important factor not only in manufacturing process but also in spring design.

### 6.2.4 Surging

A valve spring takes forced vibration by a rotating cam shaft of engine, and this vibration propagates along the wire of the spring. When the driving frequency of the cam shaft and the natural frequency of the spring coincides, there arises resonance or so-called surging. When the surging is generated, the stress of the wire increases temporarily and shortens the fatigue life. The prevention of the surging are, therefore, an important problem in the design of valve springs[19].

In the conventional dynamic analysis of springs, several assumptions are prerequisite as in the case of static analysis for a simplified calculation[20].

In FEM analysis of forced vibration, however, a valve spring can be analized as a system with degree of freedom[21, 22].

In chapter 5, the eigenvalue analysis of a valve spring was introduced. This is an analysis of nondamping free vibration and corresponds to the case of $C = P = 0$ in equation (2.32) from chapter 2. In the case of surging, however, driving vibration from the cam shaft shall be considered[23, 24]. This driving vibration is a forced vibration with a damping coefficient and corresponds to the case of $C \neq 0, P \neq 0$. In FEM analysis of surging, the method is basically the same as in the case of chapter 5, provided that the condition of forced vibration and a damping coefficient shall be added.

**(1) Analytical model** Fig. 6.46 shows an analytical model of a valve spring by eight-nodal solid elements.

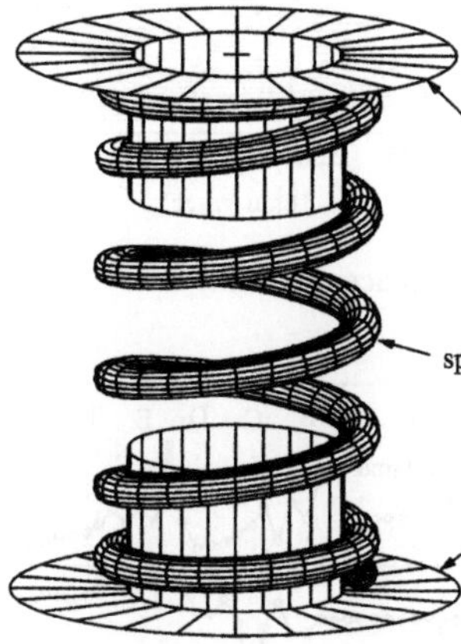


Figure 6.46: Analytical model of a valve spring.

Detail of the modeling has been explained in Paragraph 6.2.1 in which the solid element was used. The different point is that upper and lower retainer are specified as a rigid surface and the friction of interwire and friction between wire and retainer are considered. The friction coefficient is assumed to be 0.3.

The upper and the lower retainer were displaced to the height of installation and then the upper retainer was given the forced displacement following the profile of cam.

**(2) Analytical result** Fig. 6.47 and Fig. 6.48 show the time chart of individual coil displacement. Fig. 6.47 is an analytical result and Fig. 6.48 is an experimental result.

They are in good correspondence, and the propagating pattern of the surge wave

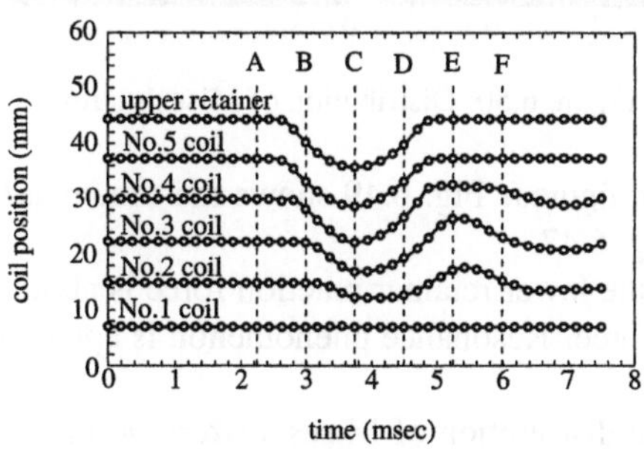


Figure 6.47: Displacement time chart. (analysis)

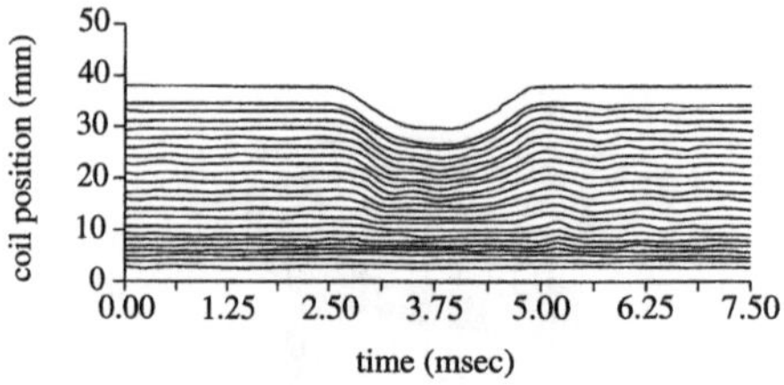


Figure 6.48: Displacement time chart. (measurement)

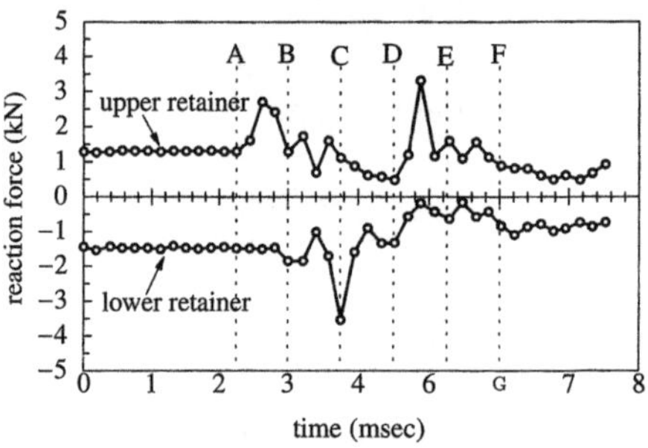


Figure 6.49: Retainer reaction force time chart.

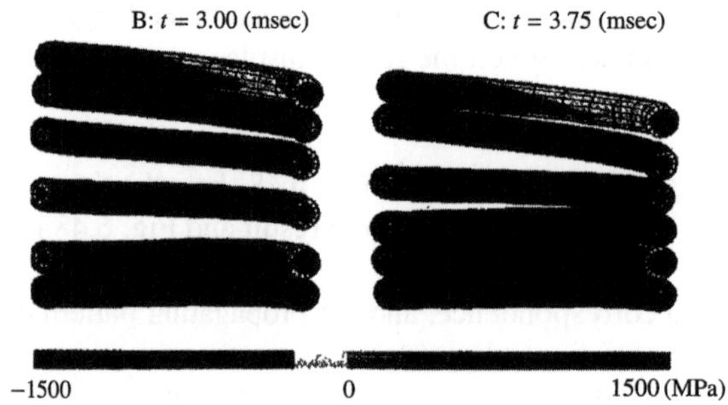


Figure 6.50: Distribution of Mises's stress.

is observable from the figures. Fig. 6.49 shows the time chart of retainer reaction force in the case of Fig. 6.47.

The peak value of the lower retainer reaction force is about two times and more than the initial static force. Resonance phenomenon is apparent from this observation.

Fig. 6.50 shows the distribution of Mises's stress on the inside diameter of the coil at step B and C in Fig. 6.47.

It can be seen that the stress at the lower side (fixed side) of the spring is higher than upper side.

## 6.3 LEAF SPRINGS

In chapter 5, the fundamental method to analyze leaf springs by FEM was explained. The following merits are explained:

- Cambered leaf can be handled.
- The analysis can be done using both symmetric and asymmetric shapes.
- A series of analyses — fastening with nip, fastening by U-bolt, and vertical loading can be treated by a unified method.

The remaining problem is contact between leaves. It will not be an exaggeration to say that the tip contact method, presented in chapter 5, is a computing method devised to simplify inter-leaf contact. Inter-leaf contact is an issue which we cannot avoid in designing leaf springs whether it belongs to a conventional type or to a progressive type. In the case of the RBA type, among others, the inter-leaf contact is especially important since the spring achieves an intense nonlinearity through its helper spring's contact along with an increased load.

When the conventional leaf spring was analyzed in chapter 5, the tip contact method gave a comparatively good approximation. For advanced calculation, however, each leaf shall be considered to contact between leaves not only at the tip and center of each leaf but also at other places. Further, there is a case in which the contact position shifts with load as in the contact between the helper spring and the main leaf in the RBA type leaf spring presented in Paragraph 5.7. In Paragraph 6.3.1, the RBA type leaf spring will be discussed.

In addition to spring itself, there are some peripheral issues to be discussed. For instance, shackles are used to install a leaf spring on a car body. The shackle has also a role to absorb the leaf span change when loaded. It is necessary to consider the large deformation to analyze the movement of the shackle. Moreover, a contact plate is used in heavy vehicle springs occasionally. When the contact plate is used, the problem of contact shall be added to the problem of large deflection. In this case, the problem of the moving-boundary condition shall be analyzed. We will discuss these issues in Paragraph 6.3.2.

A hysteresis phenomenon of leaf springs is an issue which affect the riding comfort of a car. It is caused by friction between leaves. This problem can be handled by giving a friction coefficient to contact elements. The hysteresis will be discussed in Paragraph 6.3.3.

When a vehicle, installed with leaf springs, starts or stops suddenly, a phenomenon called wind-up is caused. A rotational moment which is called a wind-up torque, besides the vertical load, comes into play on the leaf spring through the axle when the wind-up takes place. Therefore, the issue of wind-up is how to handle the axle force and moment rather than a linear or nonlinear problem. This will be discussed in Paragraph 6.3.4.

In the case of a leaf spring, presetting causes a change in camber and the length affecting the installation of the spring on the vehicle. Therefore, it is important to foresee the amount of change caused by presetting. In Paragraph 6.3.5, presetting will be discussed.

### 6.3.1 RBA Type Leaf Springs

Some leaf springs achieve a nonlinear characteristic by positively using contact[25]. The RBA type is typical of that category. Contact area between the auxiliary spring and the main spring grows when the load is increased. By this process, the overall rigidity of the spring is increased. The RBA type spring achieves a nonlinear characteristic from the contact between the auxiliary and main spring.

In chapter 5, FEM was applied to the calculation of leaf springs with fairly close results to the experimental data. This does not mean, however, all the problems of the inter-leaf contact in the leaf spring were solved. It is still difficult to handle the RBA type spring by the limited FEM technique explained thus far.

In RBA type spring, the contact pattern between the auxiliary spring and the main spring will be affected by load. To apply the tip contact method for RBA type spring, we must set up a hypothesis considering the contact pattern change with regard to the load as follows. The auxiliary spring contacts the main spring progressively and continuously starting from the center of the spring leaf[26, 27, 28, 29]. FEM enables us to remove these assumptions by inserting the contact element over the total length of each leaf[30, 31]. A unified method has been established in this manner. An analytical example will be presented here.

**(1) Analytical model** Fig. 6.51 shows RBA type spring analyzed here.

This spring is an asymmetric RBA-type spring consisting of two main leaves and one helper. Fig. 6.52 shows the analytical model of the RBA type spring in the state of center bolt fastened case and U-bolt fastened case.

The two-node two-dimensional beam element was used in modeling the main body of the RBA spring. The pad used for the fastening was modeled by the same beam element as the main body, however, the elastic coefficient was set large enough to act as a rigid element.

The contact was assumed to occur not only at the leaf tip but also on the entire length of the leaf and the contact elements were inserted between all the leaves. Constant-distance GAP elements were used as the contact element and the inter-leaf

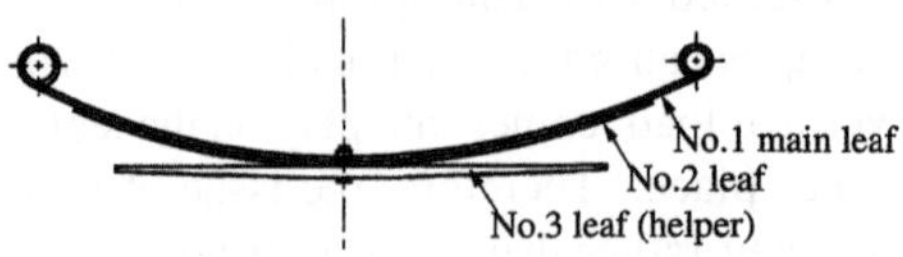


Figure 6.51: RBA type spring.

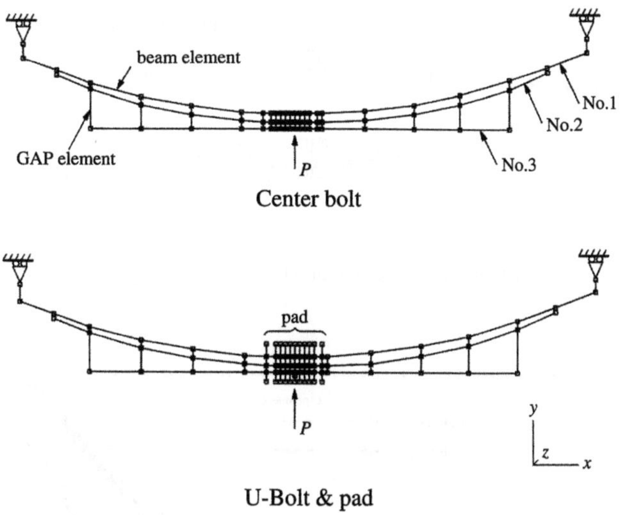


Figure 6.52: Analytical model of RBA type spring.

friction was not considered. Contact was judged by the distance between nodes, and nodes were assumed to move along the leaf after contact.

Displacement in $y$ direction of nodes at eye centers were restrained, and displacement in $x$ direction of nodes on leaf center were restrained too. Bush friction was neglected here.

The analysis was carried out for two states of the fastened and unfastened condition. The loading process was analyzed by incremental load method.

In the analysis of the fastening process by a U-bolt, the rotation of upper and lower pads was restrained keeping both pads in parallel.

**(2) Analytical result** Fig. 6.53 shows deformed state in case that a center bolt is used. Fig. 6.54 shows deformed state in case that a U-bolt is fastened.

In both cases, the helper comes in contact with No.2 leaf after deformation. The

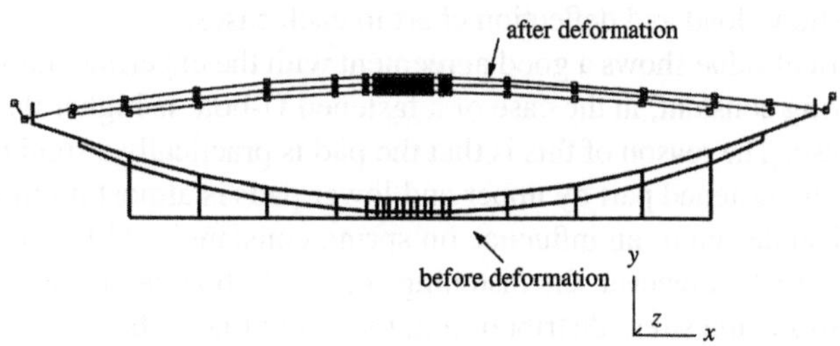


Figure 6.53: Deformed state. (Center bolt)

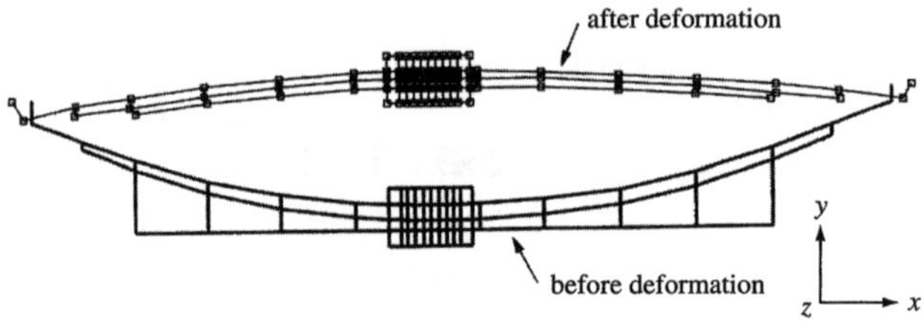


Figure 6.54: Deformed state. (U-bolt & pad)

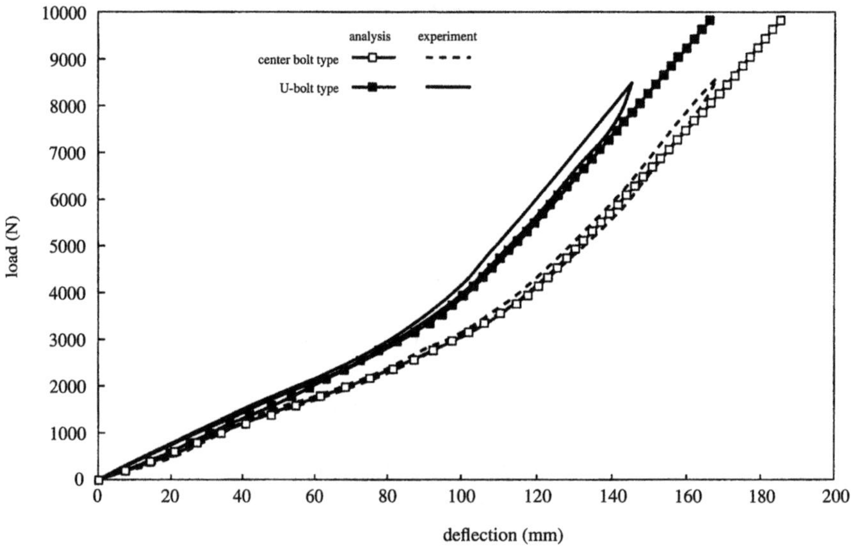


Figure 6.55: Load v.s. Deflection chart of RBA spring.

relation of load and deflection becomes nonlinear due to helper contact.

Fig. 6.55 shows load and deflection chart in each cases.

The analytical value shows a good agreement with the experimental value in each case. The spring constant, in the case of a fastened U-bolt, is higher than that of the non U-bolt case. The reason of this is that the pad is practically a rigid body and the deflection of the fastened part by upper and lower pads is almost minimal.

In a practical design of an influence on spring constant by U-bolt is assumed by experience. In FEM, however, the fastening state of U-bolt can be analyzed.

Fig. 6.56 shows the stress distribution in the case of no U-bolt.

Fig. 6.57 shows the stress distribution in the case of the U-bolt being fastened.

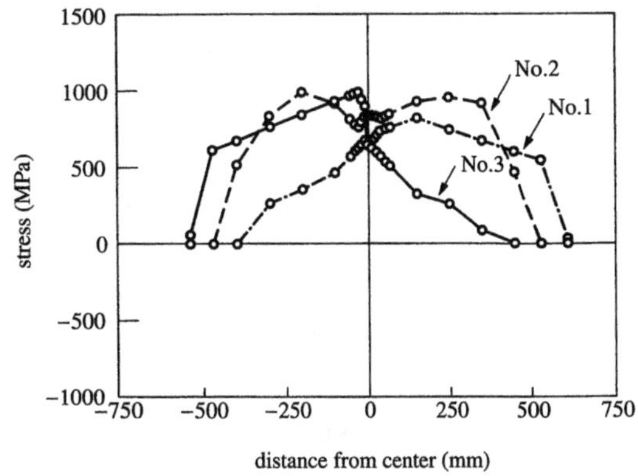


Figure 6.56: Stress distribution in RBA spring. (center bolt)

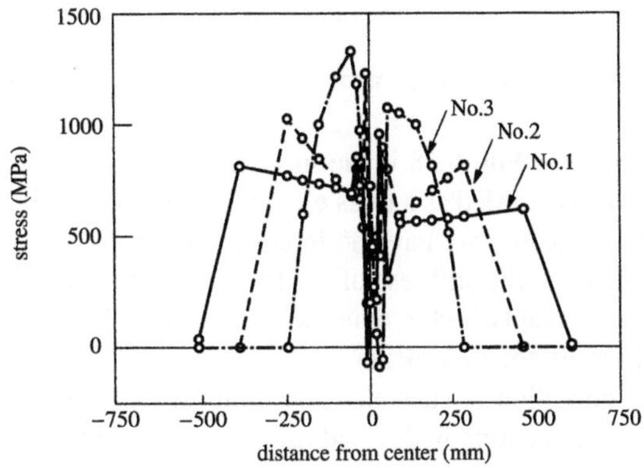


Figure 6.57: Stress distribution in RBA spring. (U-bolt)

### 6.3.2 Effect of Shackle and Contact Plate

When a leaf spring with eyes at both ends is installed in a vehicle, metal fittings called a shackle is used. One end of the shackle is connected to the eye of the spring, and the other end to the chassis. In this installation, the shackle can rotate about its coupling axis and allows the vertical or horizontal movement of the eyes within a restricted area. The span change of the spring is absorbed by this movement. The movement of the shackle affects the characteristic of a leaf spring[32, 33, 34].

In FEM, it is possible to have a shackle as a rigid element added to the analytical model of the leaf spring. In this case, however, we must solve a large deflection problem because of the shackle's large movement.

In some cases, a contact plate is installed in addition to a shackle. In this case, the leaf spring will press itself against the contact plate, and the contact point will move by load increment. Therefore, load and deflection chart is nonlinear. The analysis can also be executed by FEM, as a large deflection problem accompanied by contact.

### (1) Effect of shackle

(a) Analytical model Fig. 6.58 is an analytical model of the RBA spring with a shackle.

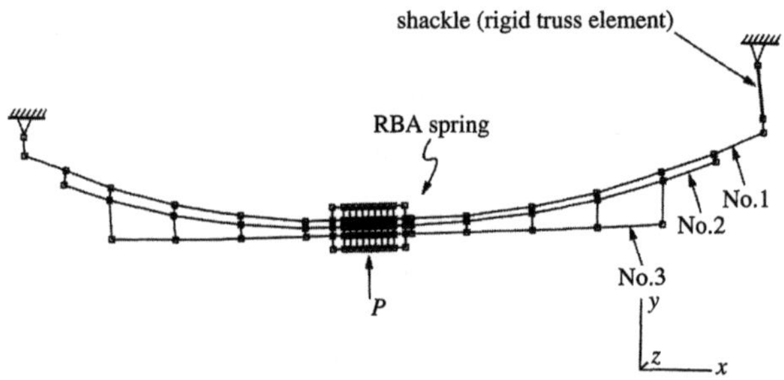


Figure 6.58: An analytical model of RBA spring with a shackle.

The important point in Fig. 6.58 is the modeling of the shackle. The shackle is installed in a way to be regarded as a truss element. The precise shape of the shackle is not necessary here because we have no intention to analyze stress in the shackle. It is necessary to presume the stiffness of the truss element. In general, the stiffness of the shackle is considerably higher than the spring itself. Therefore, the shackle is well represented by rigid truss elements.

The restraining condition is such that the connecting points to the chassis are restrained in $x$ and $y$ directions with freedom of rotation. The connecting point of spring eye and shackle is free to move.

(b) Analytical result Fig. 6.59 shows the deformed condition of the spring with a shackle.

The contact condition of the helper and main leaf after the deformation is similar

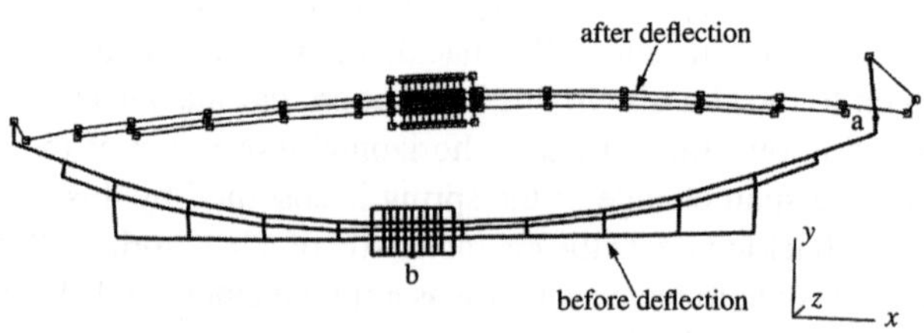


Figure 6.59: Deformed condition of the RBA spring with a shackle.

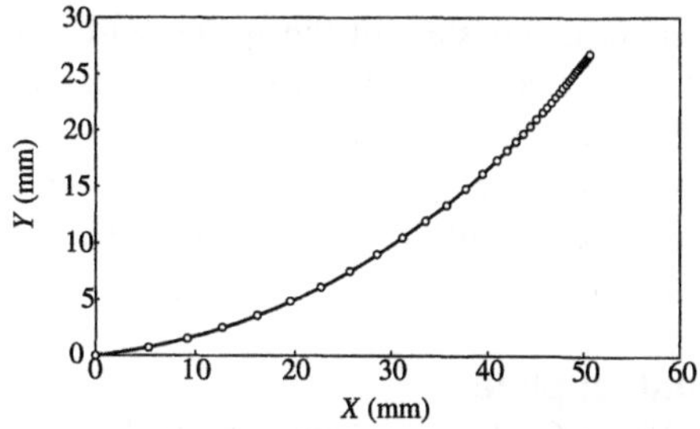


Figure 6.60: Locus of the shackle (point a in Fig. 6.59).

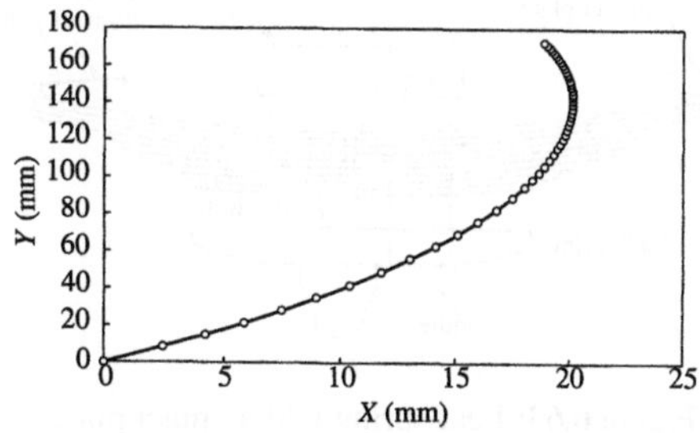


Figure 6.61: Locus of the axle center (point b in Fig. 6.59).

to the case without shackle. The locus of the eye center (point a of Fig. 6.59) in this movement is shown in Fig. 6.60.

The circular mark in the figure corresponds to a constant incremental load. Fig. 6.61 shows the locus of the axle center (point b in Fig. 6.59).

Fig. 6.62 shows load vs. deflection chart of the RBA spring with and without shackle.

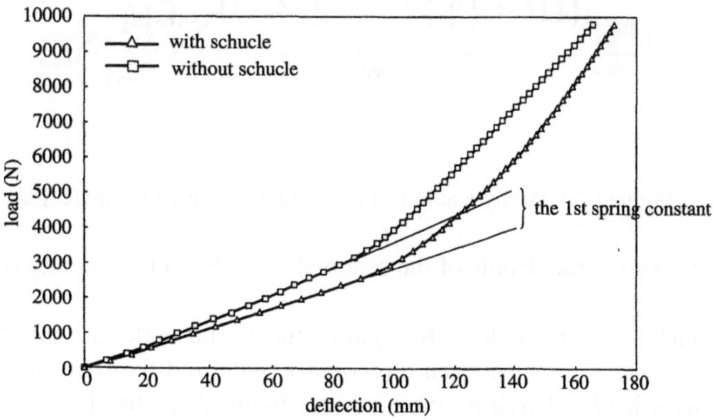


Figure 6.62: Load and deflection chart of RBA spring with and without a shackle.

Comparing these results, we can see the influence of a shackle.

For instance, the first spring constant of the spring with a shackle is lower than that of the spring without a shackle. The reason for this is that the geometrical restraint is eased by the movement of a shackle.

In the past, the movement of the shackle was traced geometrically. When FEM is used, the dynamic characteristic of the spring and the trace of the spring shackle can be treated at the same time.

### (2) Influence of contact plate

(a) Analytical model Fig. 6.63 shows a sketch of a leaf spring with contact plates.

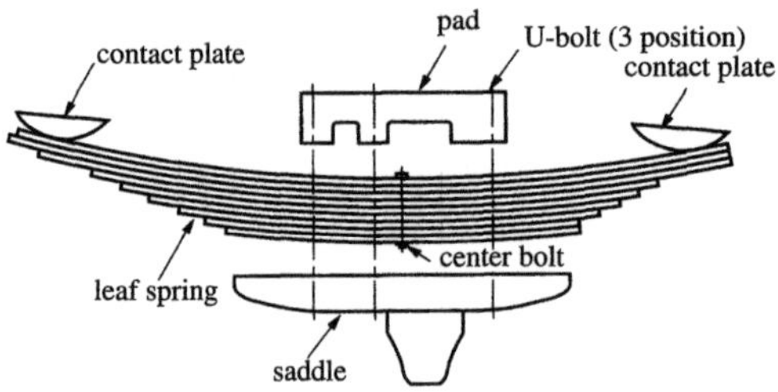


Figure 6.63: Leaf spring with contact plate.

The spring has nine main leaves and has no conventional eye. The number of U-bolts is three, differing from a conventional spring with two fastening positions. Fig. 6.64 shows the analytical model of the spring with the contact plates.

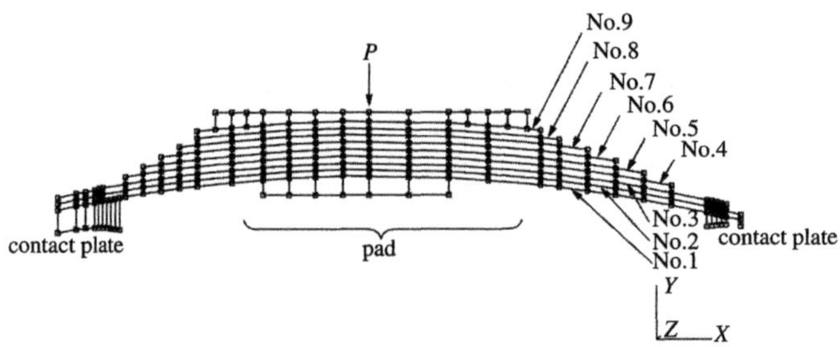


Figure 6.64: Analytical model of the leaf spring with contact plate.

Fig. 6.65 shows the details of the schematic model in the vicinity of the contact plate.

The contact plate is modeled by rigid element, and nodes are placed on its surface. The contact point between the contact plate and the main leaf changes its position with load. Therefore, the area was finely discretized and the contact element was inserted even in the area where no contact was observed at the initial condition. Friction was ignored. In the analysis, the geometrical nonlinear analysis with contact was carried out.

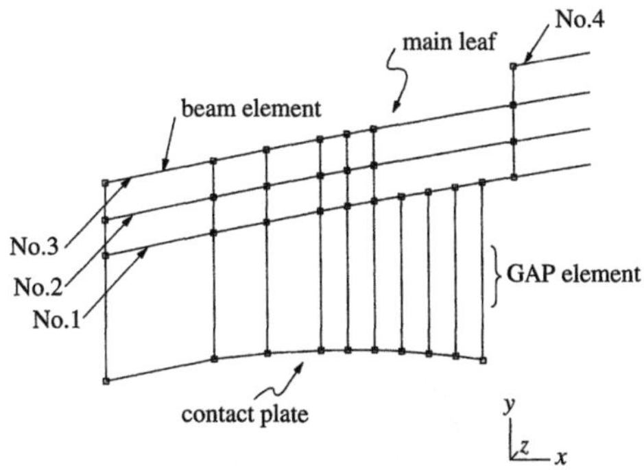


Figure 6.65: Details of the contact plate and the main leaf relation.

As to the restraining condition, the displacement in $x$ direction of all the nodes on the leaf center was restrained. The loading process during the fastening by the center bolt, the fastening by the U-bolt, and the vertical loading were given successively.

**(b) Analytical result** Fig. 6.66 shows the deformed state by a vertical load.

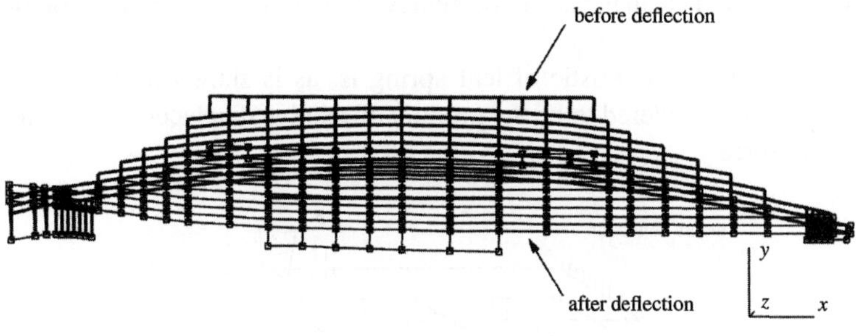


Figure 6.66: Deformed state of the leaf spring with contact plates.

It is observable that the contact position between the contact plate and No.1 leaf shifts toward the center of the leaf after deformation. Fig. 6.67 shows the relation between load and deflection.

The experimental curve shows a large hysteresis loop indicating the influence of friction. On the other hand, the analytical curve does not have any hysteresis because no friction is considered in the analysis. The steps observed in analysis curve are due to the unevenness of discretization in the area of contact plate

### 6.3.3 Hysteresis Characteristic

It has been known for a long time that a leaf spring shows a hysteresis characteristic owing to inter-leaf friction[35, 36]. The hysteresis characteristic is an important

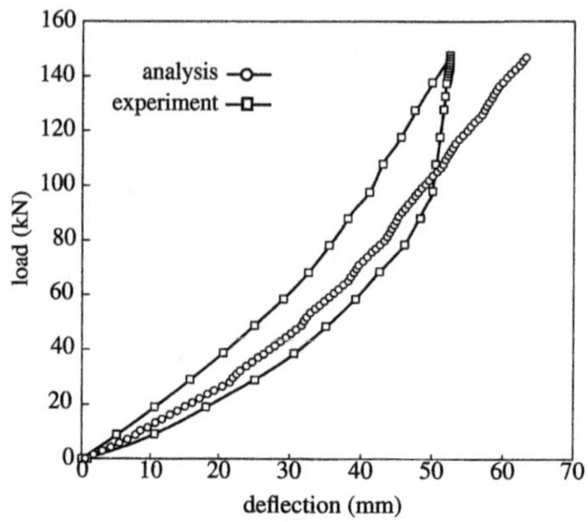


Figure 6.67: Relation between load and deflection of the leaf spring with contact plate.

factor influencing on the riding comfort of vehicles. However, the study of hysteresis characteristics is mostly limited within the experimental level, except that a few analytical reports based on tip contact method can be referred to [37, 38, 39]. In FEM, the hysteresis characteristic can be analyzed as a geometrical nonlinear problem with friction. The analysis can be achieved by giving the friction coefficient to contact elements.

The hysteresis characteristic of leaf spring is, as is shown in the conceptional figure Fig. 6.68, considered to be the product of relative displacement $\Delta S$ and inter-leaf friction force $f$.

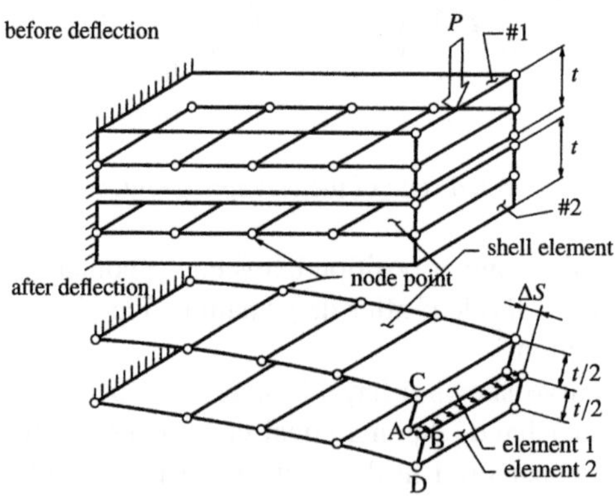


Figure 6.68: Modeling of a leaf spring by shell element.

Therefore the simulation of the relative displacement $\Delta S$ between leaves is necessary for the hysteresis curve analysis. Thickness of leaves shall be considered too. However, beam element and shell element do not encompass the factor of thickness. When we carry out nonlinear analysis, we should simulate the relative displacement of the surface of leaves. It is necessary to devise the model to evaluate $f \cdot \Delta S$. A solid element has the factor of thickness, however, the shape of the element becomes considerably long and slender, depending on the number of divisions, as is described in chapter 4, and the analytical accuracy will not be sufficient.

Hence, the shell element is used for the modeling in the present analytical example, and an improvement has even been done to simulate the relative displacement on the surface of leaves. An analytical example will be presented as follows.

**(1) Analytical model** Fig. 6.69 shows a leaf spring to be analyzed.

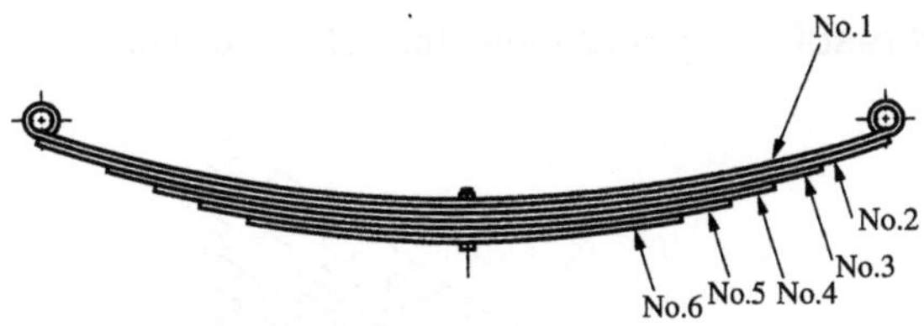


Figure 6.69: Leaf spring for analysis.

This is a conventional-type spring for a medium-sized truck. Fig. 6.70 shows the analytical model for the spring.

A quarter part of the spring was analyzed since this spring has a symmetrical shape. Fig. 6.70 shows the state before center bolt fastening. The main body of leaves are modeled by thick quadrangular shell elements. As to the inter-leaf friction, the contact at the tip and the center of the leaf was considered. To simulate thickness of the leaves, the quadrilateral thick-walled shell elements with high rigidity were put in the direction of thickness. Fig. 6.71 shows the model.

Contact elements were inserted between elements in the direction of the thickness. A constant-direction GAP element was used for the friction simulation. The friction coefficient was assumed to be 0.4.

As to the restraining condition, the displacement in $y$ direction of the eye was restrained. Nonlinear analysis with friction was carried out on the main body of the spring. A series of load conditions, the fastening by center bolt and vertical loading and unloading, was given.

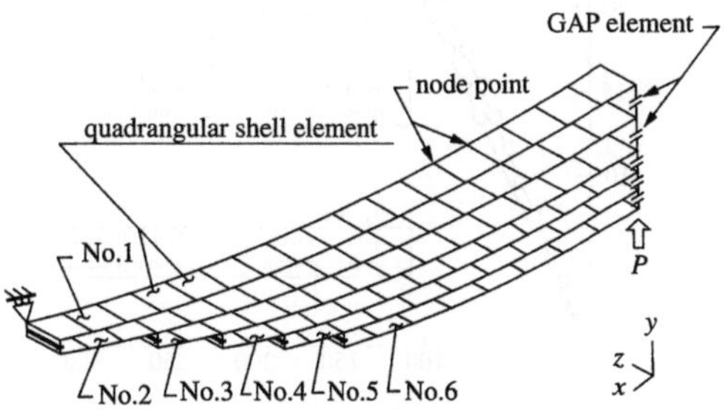


Figure 6.70: An analytical model for the spring.

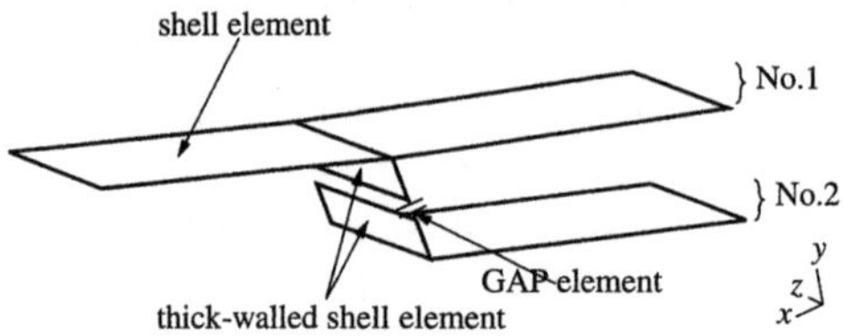


Figure 6.71: Detailed model at the tip end of the spring leaf.

**(2) Analytical result** Fig. 6.72 shows the deformed state.

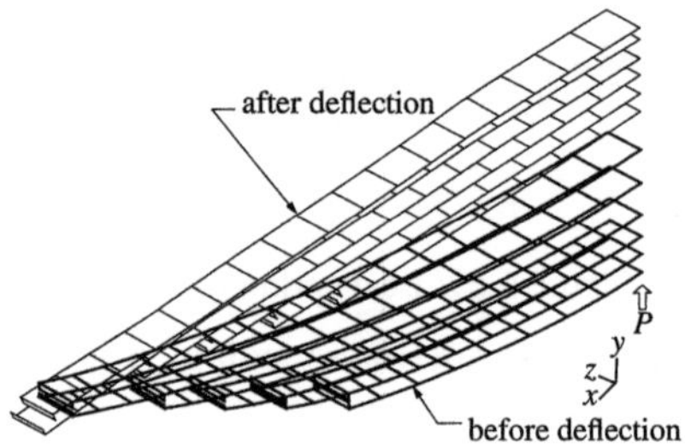


Figure 6.72: Deformed state of the leaf spring.

Relative displacement can be seen at the tip end where the contact element was inserted. The amount of relative displacement is proportional to leaf span.

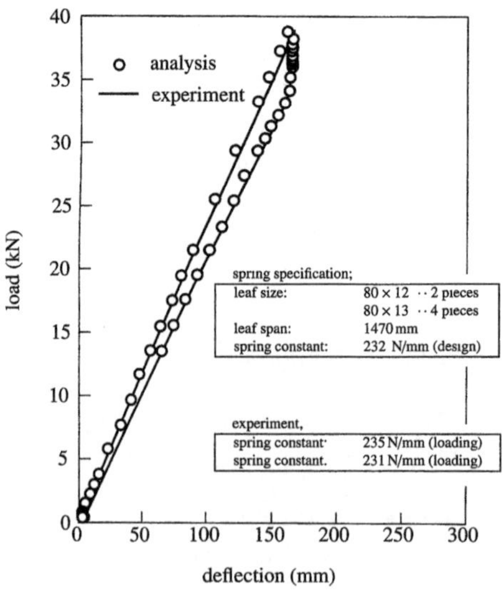


Figure 6.73: Load vs. deflection chart of the leaf spring.

Fig. 6.73 shows the load vs. deflection chart of the leaf spring.

It is observable that both the analytical and the experimental value show hysteresis. It is caused by the inter-leaf friction that is represented by the relative movement of the tip ends. The analytical and the experimental values are in good agreement.

Fig. 6.74 shows hysteresis curves for three stages of load amplitude.

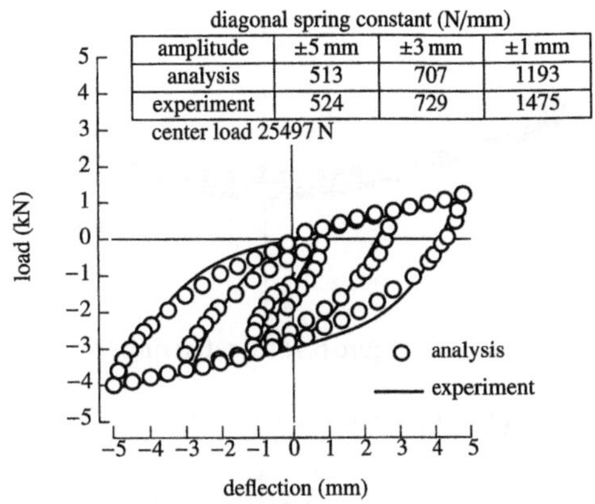


Figure 6.74: Hysteresis curves for three load amplitudes.

For each amplitude, the analytical and experimental values are in a good agreement. It seems that the FEM model construction is appropriate.

### 6.3.4 Wind-up

When a car suddenly starts or stops, front down or rear down posture will take place imposing a rotational torque "wind- up torque" to the spring. When this wind-up is generated, the spring receives not only a vertical load but also the combined action of horizontal force and rotational torque. When we analyze the wind-up by FEM, it is important to specify the combined load condition. Two methods are practicable in giving the combined load.

The first one is to have vertical load, horizontal load and rotational torque at the node of the spring directly. The problem of this method is that the selection of node on which the combined load is to be applied. Essentially, the center of rotation in which the wind-up is generated should be selected as the node. However, this node cannot be specified before analysis. Further, the center of rotation is not necessarily located at the point of node. For the combined load point, there is no other way than to have a good guess of the node point.

The second way is to construct the model of axle giving vertical and horizontal load to the load point directly. In this method, there is no need of load point prediction. The problem is how to construct the model of pad and axle. Presently, the deformation or the stress of pad and axle is not necessary. In other words, it is not necessary to faithfully model the configuration of pad and axle. They can be treated as rigid elements.

The present analysis will be carried out by the latter method.

**(1) Analytical model** Fig. 6.75 shows a leaf spring to be analyzed.

Fig. 6.76 shows an analytical model for the leaf spring.

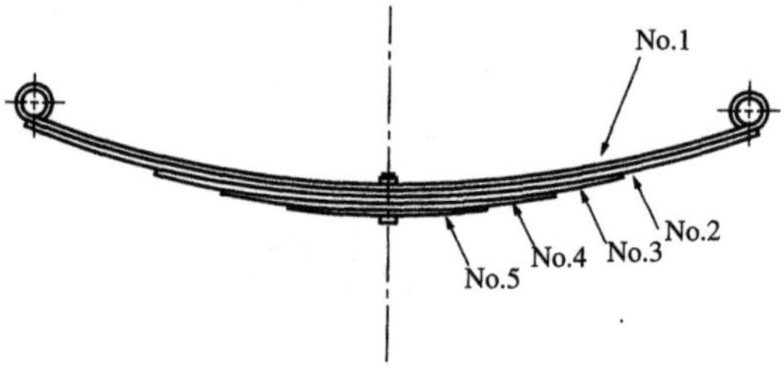


Figure 6.75: Leaf spring.

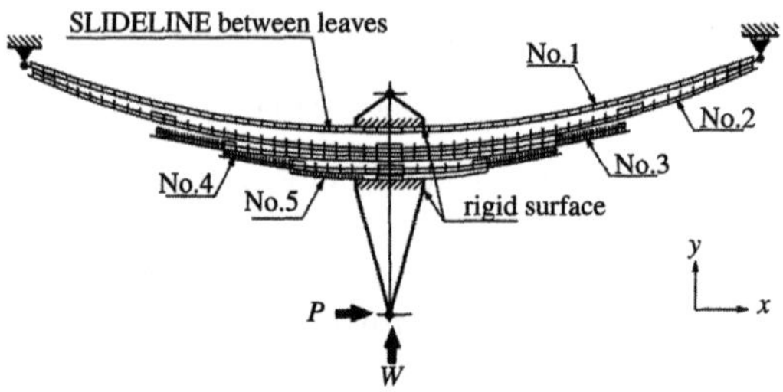


Figure 6.76: An analytical model for a leaf spring with axle.

Fig. 6.76 shows a state before fastening by center bolt. The model construction is basically the same as in the case of Fig. 6.52 with an exception that the addition of pad and axle model is added.

The stiffness of pad and axle is high enough compared with spring leaves. The pad is, therefore, defined by rigid surface, and the axle is modeled by rigid beam element as shown in Fig. 6.76. As to the inter-leaf friction, the contact nodes were applied between leaves based on modified tip-contact method. The contact surface was expressed by SLIDELINE between leaves (refer to the instruction of the general-purpose software ABAQUS). The friction coefficient was assumed to be 0.3.

The displacement in $x$ and $y$ direction at the center of front eye and $y$ direction at the center of rear eye was restrained. The load was applied to the loading point of the axle. All the center sections of the leaves were constrained in $x$ direction.

The analysis was carried out giving a series of load conditions like the fastening process by U-bolt, the loading process of vertical and horizontal load, and wind-up torque.

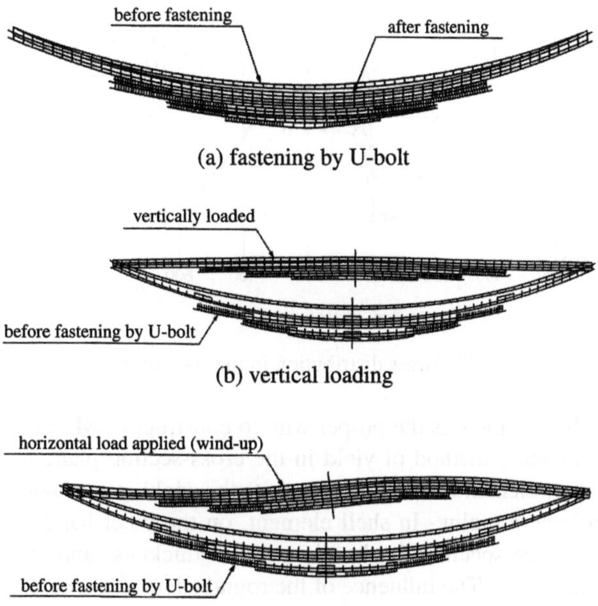


(a) fastening by U-bolt

(b) vertical loading

(c) wind-up process

Figure 6.77: Deformed state of the leaf spring.

**(2) Analytical result** Fig. 6.77(a) shows the deformed state after U-bolt was fastened.

Originally, the spring leaves are fastened by center bolt before U-bolt, and the stress is generated in this process which is called nip stress. As was explained already in Chapter 5, the nip stress (compressive stress) is generated on the upper surface of main leaf, and offsets tension stress caused by load.

Fig. 6.77(b) shows the deformed state of the spring when the vertical load was applied after fastening by U-bolt. Fig. 6.77(c) shows the deformed state of the spring when horizontal load is additionally applied at the loading point in the state of (b). Fig. 6.77(c) presents the deformed shape of spring in the wind-up state. We see in Fig. 6.77(c) that the leaf spring is deformed into the shape of letter S as a whole under the action of wind-up torque. Fig. 6.78 shows the stress distribution generated in each leaf when the wind-up torque is applied on the spring.

The analytical value shows a comparatively good agreement with the experiment. Calculated values are high at the center part of No.1 and No.5 leaf. This could be attributed to an inadequate modeling of the connection part of leaf and center bolt.

### 6.3.5 Presetting

Presetting of leaf springs can be treated as an elasto-plastic analysis.

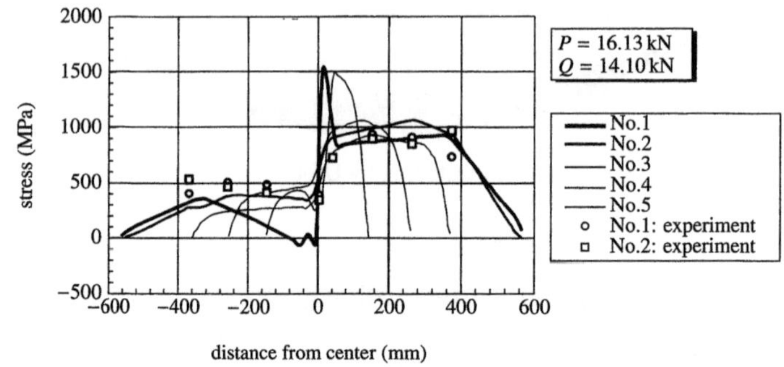


Figure 6.78: Stress distribution in the state of wind-up.

A beam or shell element is the proper way to construct FEM modeling of a leaf spring. The evaluating method of yield in the cross-section plane is not common in beam and shell element. In beam element, the yield is evaluated by stress at predetermined integral point. In shell element, on the other hand, the number of divided layers can be specified in the direction of thickness and the yield can be evaluated for each layer. The influence of the round edge of leaf cannot be included in both elements. In treating an arbitrary cross-section, solid element should be used instead. In the present analysis, however, beam element method will be explained.

**(1) Analytical model** Fig. 6.79 shows a leaf spring to be analyzed.

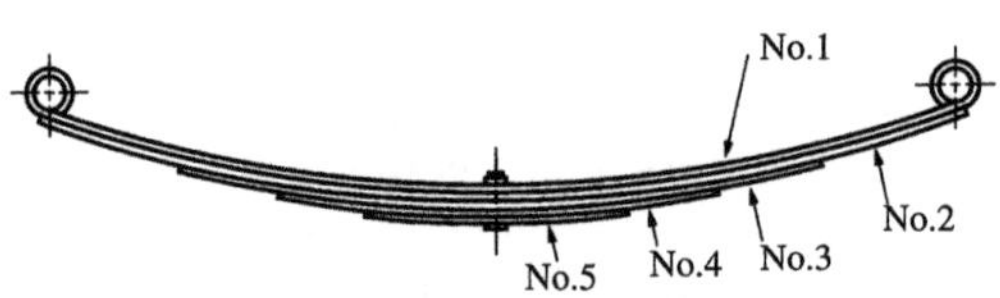


Figure 6.79: Leaf spring.

Fig. 6.80 shows the analytical model of the leaf spring.

It shows the spring in the state before fastening by center bolt. For the modeling of the spring, two-node two-dimensional beam elements were used. The method of modeling is basically similar to the case in subsection 6.3.4.

An elasto-plastic analysis will be employed. When beam element is used for the elasto-plastic analysis, the judgment of yield depends on stress at the integral point of element, as is described in the case of helical spring in subsection 6.2.3. Fig. 6.81 shows the integral points of two-node two-dimensional beam element.

As to inter-leaf friction, a contact node was set between leaves based on modified tip-contact method. The contact was expressed by interface and slide line. Friction coefficient was assumed to be 0.3.

As to restraining condition, $x$ and $y$ displacement at the center of both eyes of main leaf was restrained. All the nodes at the center of each leaf were constrained

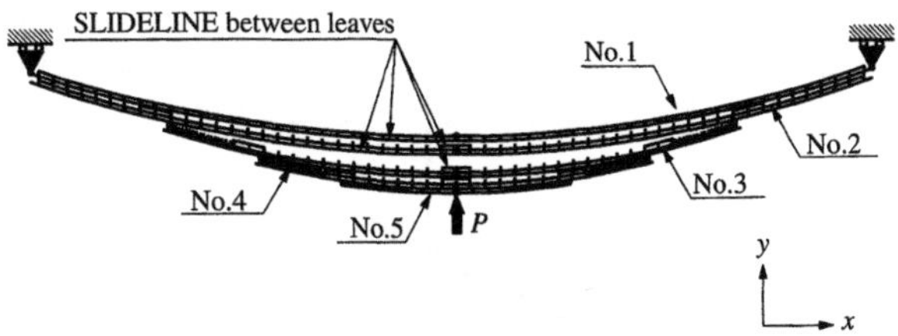


Figure 6.80: Analytical model of leaf spring.

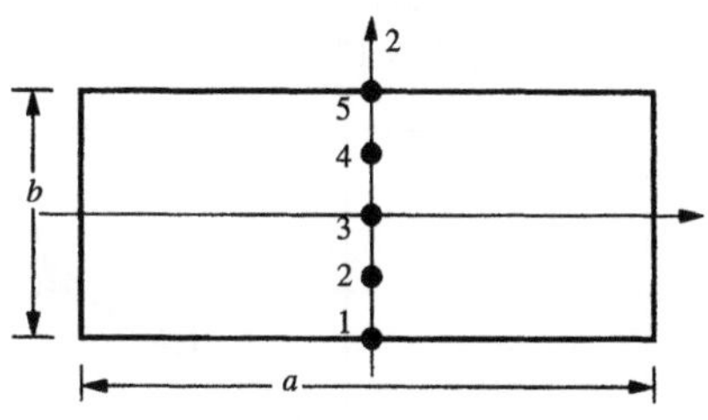


Figure 6.81: Integral points of two-node two-dimensional element.

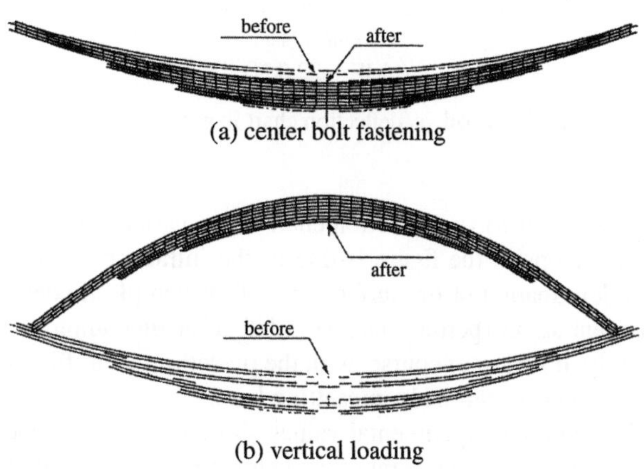


Figure 6.82: Deformed state of leaf spring (in the state of presetting).

in $x$ direction. The analysis was carried out giving a series of load conditions like center bolt fastening and vertical loading.

**(2) Analytical result** Fig. 6.82 shows the deformed state of the leaf spring.

Fig. 6.82(a) shows the deformed state after center bolt fastening, and (b) shows the state after vertical loading, that is, in the state of presetting. The vertical load (presetting load) was applied so that a part of the surface of leaf is deformed plasti-

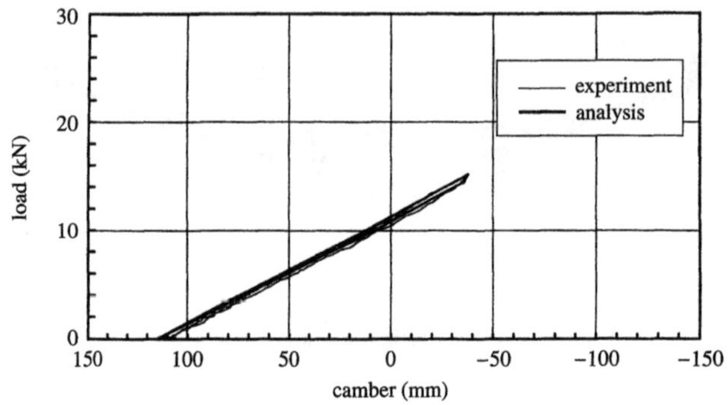


(a) $W = 15$ (kN)

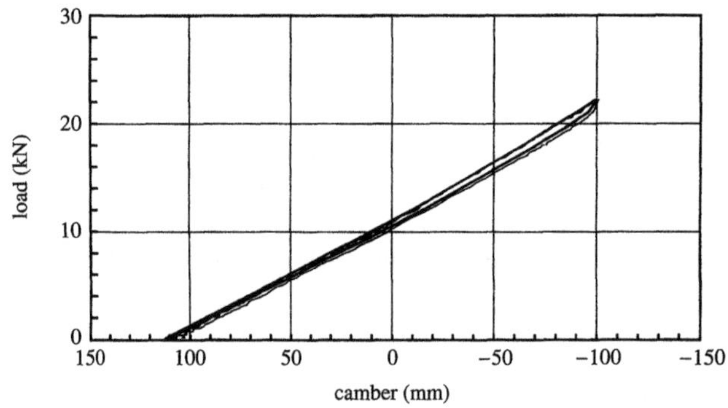


(b) $W = 22$ (kN)

Figure 6.83: Load vs. deflection chart in presetting process.

cally.

Fig. 6.83 shows the load vs. deflection chart under various presetting loads.

The hysteresis loops in the figure is due to the influence of inter-leaf friction, and the plastic deformation of the surface of leaf. When plastic deformation takes place, a permanent strain (permanent set) is left after unloading, and the camber varies accordingly. It varies, of course, with the presetting load. Fig. 6.84 shows the increase of permanent set caused by presetting load.

The analytical and the experimental values show a good correspondence. The amount of permanent set increases rapidly when the presetting load exceeds 20 kN. The amount of camber change caused by presetting is predictable by FEM analysis as we have seen thus far. It is important in the design of multi-leaf spring to foresee the amount of camber change.

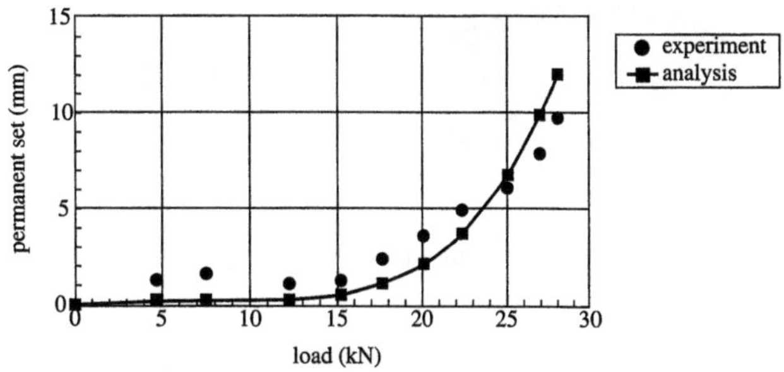


Figure 6.84: Relation between presetting load and permanent set.

## 6.4 DISC SPRINGS, RING SPRINGS

These springs can be treated as an axisymmetric analysis. An arbitrary cross section can be handled easily. Hence, calculation formulas for them have been established with good approximation from experiments. Merit by applying FEM is to be able to do a fine examination. Some examples will be presented in this section.

### 6.4.1 Disc Springs

Almen's equation is used widely as a design formula for disc springs[40]. This formula is an analysis taking geometrical nonlinearity into account with a good agreement to experiment. The formula can be interpreted, from the standpoint of FEM[41], that the cross-section is replaced by a single axisymmetric shell element. The application of Almen's formula has limitation as follows:

- It cannot handle a chamfer.
- It cannot consider the influence of friction.

An analytical example is shown below:

**(1) Analytical model** Although there is a disk spring with a slit[42], a plain disc spring without slit will be discussed here.

In addition to a single use, a combined use like in series or parallel as is shown in Fig. 6.85 is a common way of disc spring setting.

Fig. 6.86 shows the analytical model for each type of configuration.

Axisymmetric element modeling is appropriate for a disk spring. Eight-node quadrilateral axisymmetric elements were used here to consider the roundness of chamfered section. Between contact surfaces, joining elements suitable for the axisymmetric element were inserted. Friction coefficient was assumed to be 0.3.

As to the restraining condition, the plane a-a and the plane b-b were defined as rigid surfaces and the plane b-b was restrained completely. Lateral and rotational

(a) series (b) parallel

Figure 6.85: Combination of disk springs.

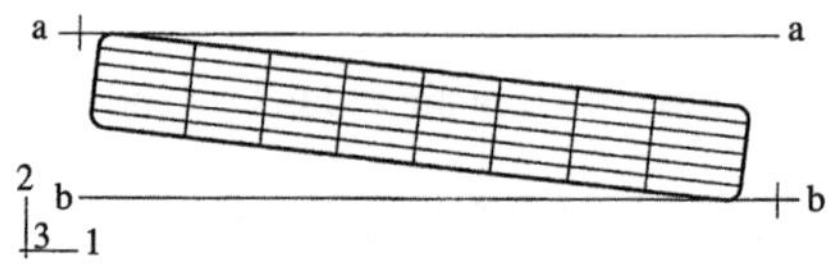


(a) single piece

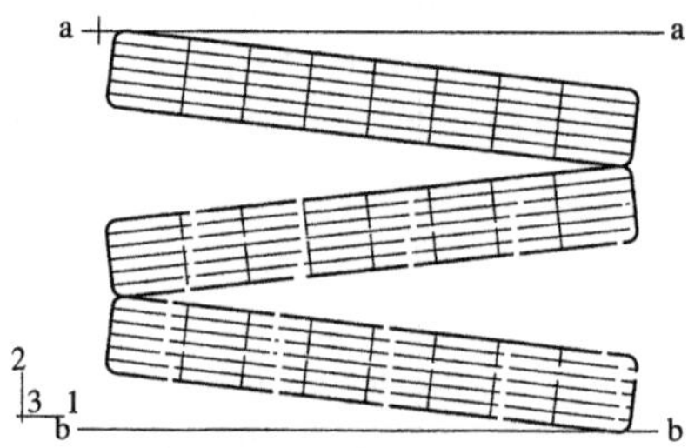


(b) series pile

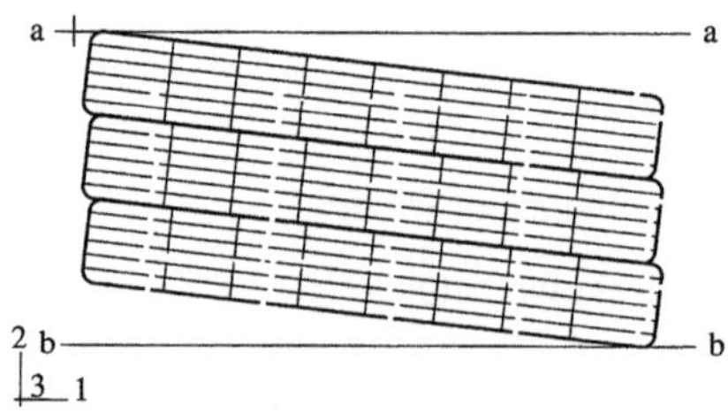


(c) parallel pile

Figure 6.86: Analytical models for the cross-section of disc spring in each combination.

displacement of the plane a-a were restrained and the forced displacement (the displacement corresponding to loading and unloading) was given in vertical direction. Contact between rigid surfaces (plane a-a or plane b-b) and the disc spring and, also between disc springs themselves were considered. Besides nonlinear analysis with friction, linear analysis was also carried out as a reference.

**(2) Analytical result** Fig. 6.87 shows the deformed state for each type of combination.

Fig. 6.88 shows load vs. deflection chart comparing the result of experiment, linear analysis, and non-linear analysis.

The experimental value is in good agreement with non-linear analysis including the tendency of hysteresis. The hysteresis loop area depends on the combination type. Parallel combination is larger than series type. Apparently, larger contact surface contributes to greater hysteresis. In nonlinear analysis, there is a point where load increases rapidly. This shows that the surface of disc spring is almost in contact with rigid surface at this stage.

Fig. 6.89 shows the distribution of principal stress in the cross-section of disc spring.

It is seen that tensile stress is generated on the lower surface, and compressive stress on the upper surface.

As you see, FEM analysis is indispensable in the analysis of stress distribution

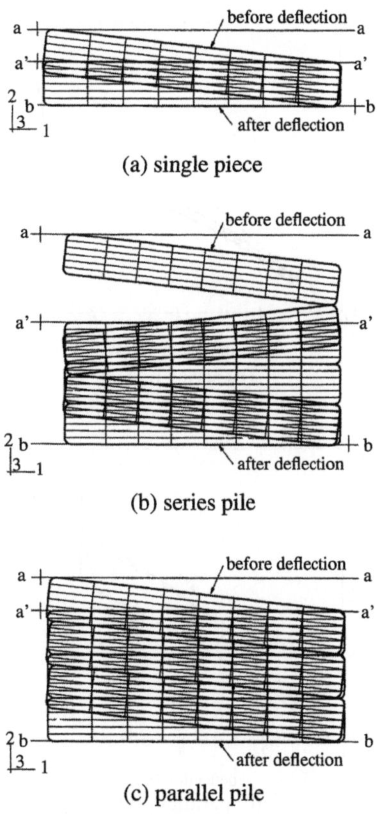


Figure 6.87: Deformed state of disc spring.

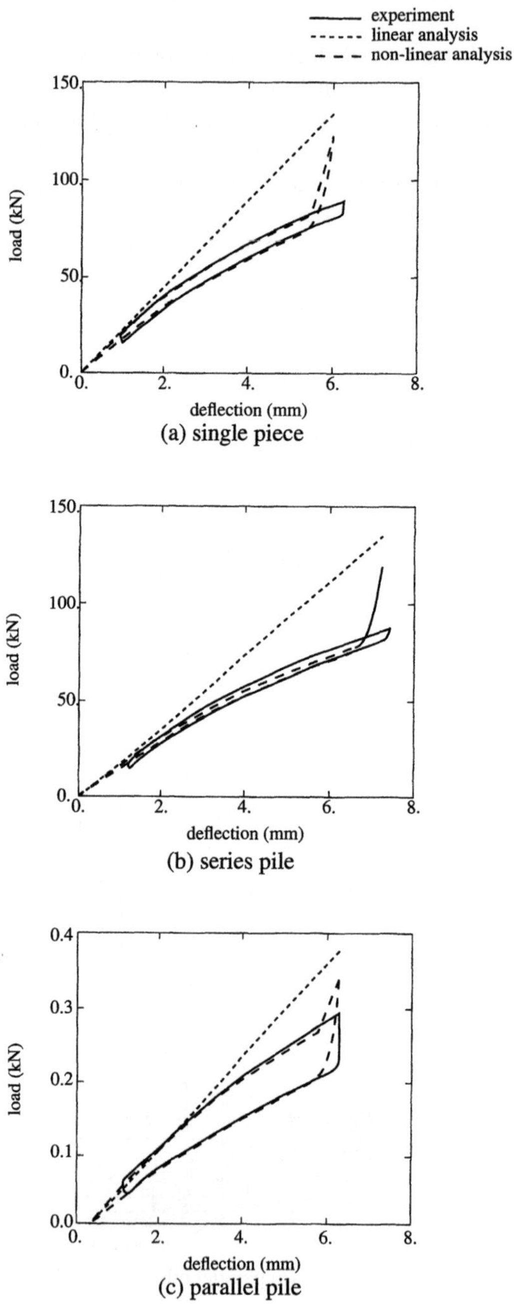


Figure 6.88: The load vs. deflection chart.

Figure 6.89: Distribution of principal stress in the cross-section of disc spring.

in cross-section. In the use of disc spring, load characteristic could be controled easily without changing its dimensions. This can be possible through the adjustment of residual stress by presetting. Such an elasto-plastic analysis involving process simulation can be done by FEM analysis[41].

### 6.4.2 Ring Springs

The design formula for ring springs[43] is based on the following assumptions on mechanics:

- deformation is assumed to be minute;
- the cross section of ring is pentagonal;
- stress within the ring is only hoop stress; and,
- stress within the ring is constant.

In the case that those assumption are not adequate, it is FEM's turn to be pursued. Even if the actual shape of the cross-section is hexagonal as is exemplified in Fig. 6.90, we have to approximate it by a pentagon. Stress concentration or axial stress cannot be a target of analysis in this conventional design formula.

**(1) Analytical model** Fig. 6.90 shows a rough sketch of a ring spring. Fig. 6.91 shows an analytical model of the cross-section of ring spring.

In Fig. 6.91, a part of the cross section (a shaded part in Fig. 6.90) was selected considering axial symmetry. As no curve was included in the cross section, four-node tetrahedral axisymmetric elements were used in modeling.

The contact element was inserted in the contact area (on c-c and c'-c') between outer and inner rings. To consider friction using contact element, it is required to specify friction coefficient.

As to the restraining condition, the displacement in $x$ direction of nodes on b-b

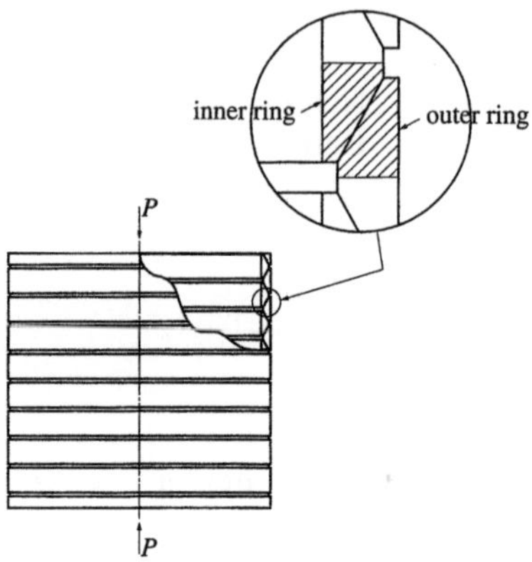


Figure 6.90: Ring spring.

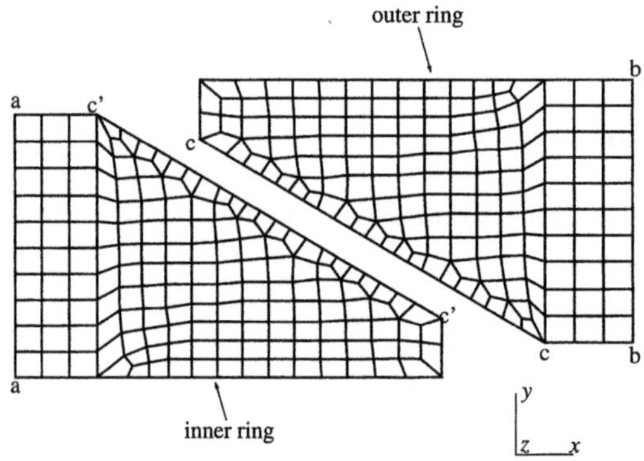


Figure 6.91: Analytical model of the cross-section of ring spring.

was restrained and load (loading and unloading) was applied to nodes on a-a in $x$ direction. The restraint on the displacement of outer or inner ring in $y$ direction was not necessary because the element was axisymmetric. The FEM calculation was executed by nonlinear analysis with friction. The friction coefficient was assumed to be in two levels: 0.029 and 0.29.

**(2) Analytical result** Fig. 6.92 shows the relation between load and deflection.

In any case of (a) and (b), the relation between load and deflection shows hysteresis. In Fig. 6.92, case (b) is larger than case (a) as to energy absorption due to larger friction coefficient.

Fig. 6.93 and Fig. 6.94 show the stress distribution at maximum load.

Fig. 6.93 shows the distribution of hoop stress on the cross-section of ring spring. Compressive stress is generated in inner ring and tensile stress in outer ring. Fig. 6.94 shows the distribution of Mise's stress.

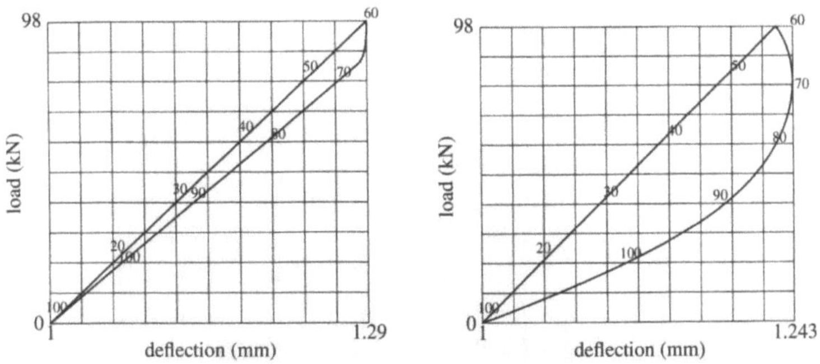


(a) friction coefficient 0.029 (b) friction coefficient 0.29

Figure 6.92: Load vs. deflection chart of ring spring.

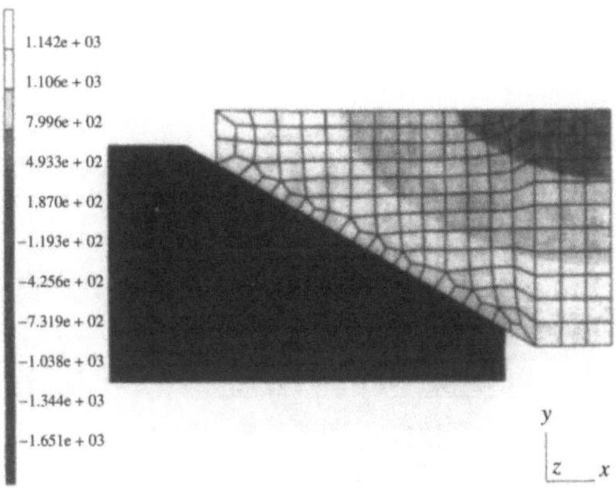


(a) friction coefficient 0.029

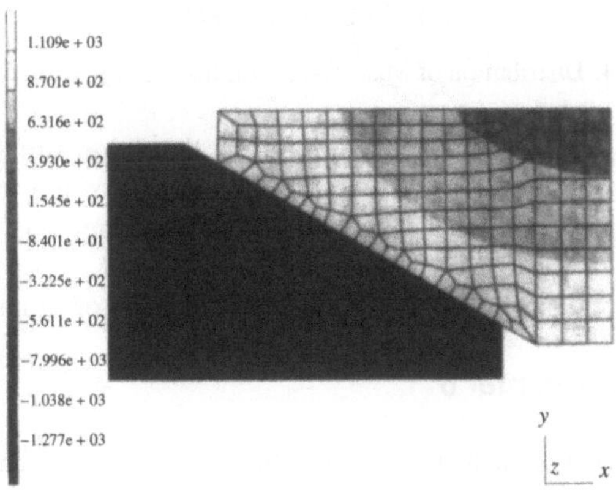


(b) friction coefficient 0.29

Figure 6.93: Distribution of hoop stress on the cross-section of ring spring.

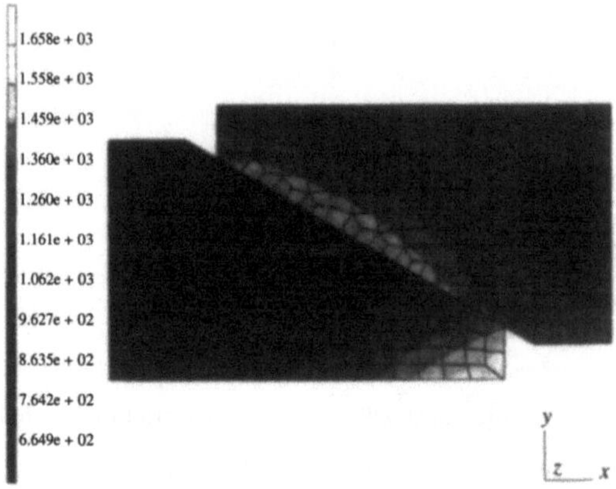


(a) friction coefficient 0.029

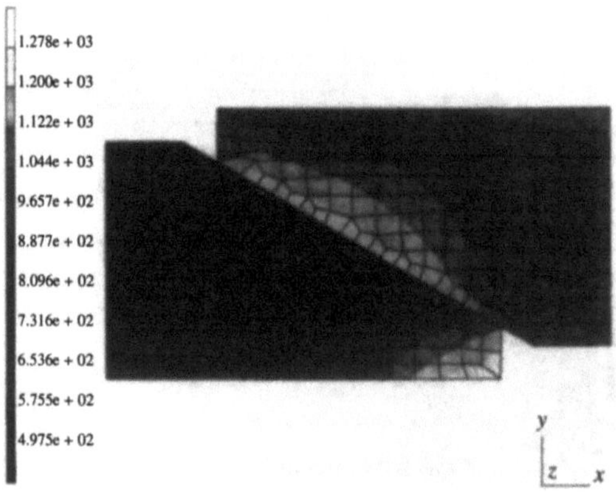


(b) friction coefficient 0.29

Figure 6.94: Distribution of Mise's stress on the cross-section of ring spring.

It is observable from Fig. 6.93 and Fig. 6.94 that hoop stress and Mise's stress show a similar pattern of stress distribution regardless of the friction coefficients. Although an axial stress distribution was not shown here, it is a part of the FEM analysis too.

Thus, FEM is indispensable for the stress distribution of ring spring.

## Bibliography of Chapter 6

[1] Inoue K, Design of the hollow stabilizer (in Japanese), Transactions of JSSR (1983), p46
[2] Satoh N, Ogawa T, Hamano T, FEM Analysis of Hollow Stabilizer with Flat Section (in Japanese), Transactions of JSSR (1996), p27
[3] JSSR, Bane 3rd ed. (in Japanese) 1982, Maruzen, p214

[4] Watari A, Kobayashi S, Design of Variable Pitch Spring (in Japanese), Transactions of JSSR (1959), p93
[5] Kobayashi S, Design of Coil Springs with Variable coil diameter (in Japanese), Transactions of JSSR (1959), p97
[6] Oguchi S, Design of Conical Coil Springs (in Japanese), Transactions of JSSR (1973), p85
[7] Suzuki K, Shimoseki M, On the Non-linear Coil Springs (in Japanese), Transactions of JSSR (1974), p47
[8] Iizuka T, Automotive Conical Taper Coil Springs (in Japanese), Transactions of JSSR (1981), p35
[9] Inoue T, Higuchi J, On the Barrel Taper Coil Spring (in Japanese), Transactions of JSSR (1981), p40
[10] Shimoseki M, Sadakuni Y, Analysis of Coil Springs using Joint Elements (in Japanese), Transactions of JSSR (1987), p1
[11] Hamano T, Tachikawa T,Analysis of compression coil springs by finite element method with joint elements (in Japanese), Transactions of JSSR (1993), p25
[12] Sato M, Matsumoto Y, Saito J, Morita K, Stress Analysis of Coil Springs with Arbitrary Cross Section (in Japanese), Transactions of JSSR (1986), p13
[13] JSSR, Bane 3rd ed. (in Japanese) 1982, Maruzen, p192
[14] Nagaya K, Stress and Vibration Analysis of a Coil Spring of General Shape having an Arbitrary Shaped Cross Section (in Japanese), Transactions of JSSR (1988), p3
[15] Ohkochi T, Imaizumi T, Ichikawa S, Shape Optimization of the Wire Cross Section of a Helical Spring (in Japanese), Transactions of JSSR (1991), p23
[16] Shimoseki M, Fujinuma H, Chino H, An Isoparametric Element for curved and twisted beam with Rectangular Section (in Japanese), Transactions of JSSR (1994), p33
[17] Shimoseki M, Oshida M, Variation of Height due to Pre-setting in Coil Springs (in Japanese), Transactions of JSSR (1975), p70
[18] Shimoseki M, Sato M, Inference of Change of Height due to Pre-setting in Coil Springs with Varied Pitch and Coil Diameter (in Japanese), Transactions of JSSR (1980), p46
[19] Kato M, Vibration Analysis of Conical Springs (in Japanese), Transactions of JSSR (1976), p18
[20] JSSR, Bane 3rd ed. (in Japanese) 1982, Maruzen, p209
[21] Sadakuni Y, Shimoseki M, Dynamic Analysis of Coil Springs by Finite Element Method (in Japanese), Transactions of JSSR (1987), p11
[22] Sawado K, Oh K, Computerized Symbolic Manipulation in Analysis of Characteristic of Helical Spring (in Japanese), Transactions of JSSR (1989), p18
[23] Sadakuni Y, Shimoseki M, Dynamic Response Analysis of Coil Springs by Finite Element Method (in Japanese), Transactions of JSSR (1988), p21
[24] Nagaya K, Design Method for Coil Springs of Arbitrary Shape with Arbitrary Wire Cross Section under General Transient Excitation (in Japanese), Transactions of JSSR (1989), p8
[25] JSSR, Bane 3rd ed. (in Japanese) 1982, Maruzen, p214
[26] Kobayashi S, Design Formula of Progressive Spring 1st Report (in Japanese), Transactions of JSSR (1955), p52
[27] Kobayashi S, Design Formula of Progressive Spring 2nd Report (in Japanese), Transactions of JSSR (1955), p55
[28] Kobayashi S, Design Formula of Progressive Leaf Spring by complete contact model (in Japanese), Transactions of JSSR (1959), p58

[29] Aoyama S, Yoshikawa K, Calculation of Stress Distribution of Conventional Leaf Springs for Motor Truck Use (in Japanese), Transactions of JSSR (1977), p43
[30] Shimoseki M, Hayasaka Y, Analysis of Leaf springs by RBSM Method (in Japanese), Transactions of JSSR (1984), p12
[31] Hamano T, Itou M, Analysis of leaf spring using finite element method with a joint element (in Japanese), Transactions of JSSR (1991), p39
[32] Kobayashi S, Load vs. Deflection Characteristics of Leaf Springs with a Shackle (in Japanese), Transactions of JSSR (1959), p61
[33] Chisima K, Various Load Characteristics of Leaf Springs with a Shackle (in Japanese), Transactions of JSSR (1962), p124
[34] Chisima K, Stress of Leaf Springs with a Shackle (in Japanese), Transactions of JSSR (1962), p131
[35] Nagashima K, Saga J, On the Hysteresis curve of the Laminated Spring (in Japanese), Transactions of JSSR (1952), p62
[36] Saga J, Deflection of Laminated Spring (Finding Hysteresis curve 2nd Report) (in Japanese), Transactions of JSSR (1955), p28
[37] Yamamoto T, Ota H, The Effects of Interleaf Friction of Laminated Leaf Springs on Their Vibration Characteristics (1st Report, Response Curves of Vibratory System with Leaf Spring), (in Japanese), Transactions of JSSR (1973), p28
[38] Ota H, Sugiura F, Kawasaki M, The Effects of Interleaf Friction of Laminated Leaf Springs on Their Vibration Characteristics (2nd Report, Comparison between Static and Dynamic Hysteresis curve), (in Japanese), Transactions of JSSR (1973), p39
[39] Ota H, Ishida T, The Effects of Interleaf Friction of Laminated Leaf Springs on Their Vibration Characteristics (3rd Report, Dynamic Properties of Model Springs with Arbitrary Shape, Dimension, and Leaves), (in Japanese), Transactions of JSSR (1973), p49
[40] JSSR, Bane 3rd ed. (in Japanese) 1982, Maruzen, p283
[41] Shimoseki M, Hayasaka Y, Analysis of Disk springs (in Japanese), Technical Report of JSSR Meeting (1983 Fall), p41
[42] Aoyama S, Yoshikawa K, Deflection and State of Stress on the Coned Disk of Clutch Diaphragm Spring (in Japanese), Transactions of JSSR(1973), p77
[43] JSSR, Bane 3rd ed. (in Japanese) 1982, Maruzen, p304

# Index

MIX
Papier aus verantwortungsvollen Quellen
Paper from responsible sources
FSC® C105338

If you have any concerns about our products,
you can contact us on
**ProductSafety@springernature.com**

In case Publisher is established outside the EU,
the EU authorized representative is:
**Springer Nature Customer Service Center GmbH**
**Europaplatz 3, 69115 Heidelberg, Germany**

Printed by Libri Plureos GmbH
in Hamburg, Germany